Food for Africa
The life and work of a scientist in GM crops

Food for Africa
The life and work of a scientist in GM crops

Jennifer Thomson

UCT
P R E S S

Food for Africa: The life and work of a scientist in GM crops

First published 2013
by UCT Press
an imprint of Juta & Company Ltd
First floor
Sunclare Building
21 Dreyer Street
Claremont
7708
South Africa

© 2013 UCT Press
www.uctpress.co.za

ISBN 978-1-92049-981-5
ePub 978-1-77582-049-9
WebPDF 978-1-77582-048-2

Project manager: Karen Froneman
Editor: Leonie Hofmeyr-Juritz
Proofreader: Lee-Ann Ashcroft
Cover designer: Monique Cleghorn
Typesetter: Integra Software Service Pvt. Ltd
Printed in South Africa by XXX

Typeset in Century Regular 10/13

This book has been independently peer reviewed by academics who are experts in the field.

Contents

Foreword

I am a woman of science, having studied medicine and qualified as a doctor, as well as achieving diplomas in tropical medicine and hygiene, and in public health. And I am actively interested in the health of Africans. During my anti-apartheid activist years in the 1970s I was involved in organising and working with community development programmes like the Zanempilo Community Health Centre in Zinyoka (a village outside King William's Town in the Eastern Cape), one of the first primary healthcare initiatives which was outside the public sector in South Africa. Banished by the apartheid government in 1977 to a remote town in the north, Tzaneen, I formed the Isutheng Community Health Programme, where one of the projects I encouraged was one in which women established vegetable gardens. So I have seen at grassroots level what struggles the rural poor face in getting access to enough food. And as a managing director of the World Bank I have helped create programmes that enable the poor in developing countries to shape a better future for themselves.

Jennifer Thomson's life as a woman scientist, as told in this book, is one with which I can identify. So is her journey to the top rungs of the ladder from which she has addressed the world's leaders. And her remarkable role in the evolution of her chosen field should be celebrated. But it is not just as a laboratory scientist in a white coat bent over her microscope that she has made her mark. She has also thrown herself wholeheartedly into assisting government in developing policies for a technological future.

I deeply appreciate the issues of food security that confront Africans. Programmes that involve and assist smallholder farmers to provide food sustainably, for themselves and their communities, should be supported. And Professor Thomson's enthusiastic efforts outside the laboratory to mobilise governments and big business have resulted in them financing the experiments and projects that are leading to the creation of more abundant and reliable crops. She is indeed a role model—for women as well as budding scientists.

The work of Jennifer Thomson has been transformational in demonstrating the value of biotechnology to food security on a continent that suffers from droughts and adverse weather patterns.

The suspicions around the African continent towards scientific interventions can only be dispelled by work such as she is championing to make science work for society.

Dr Mamphela Ramphele
Chair of the Technical Innovation Agency (TIA)

Acknowledgements

Every author needs friends who will read draft manuscripts with a critical and insightful eye. I was fortunate in having Lesley Shackleton, with whom I co-founded South African Women in Science and Engineering. She impressed on me the importance of showing the pivotal role that women had played in my story. I also had Nancy van Schaik, my former head of department at Wits University. She pointed out major gaps in my recollections and flaws in my memory. Lynne du Toit, CEO of Juta, saw a glimmer of possibility in my idea for this book and handed me over to Sandy Shepherd at UCT Press who had to struggle with a very messy first draft. Thank you Sandy for sticking with it!

Among the many people I talk about in this book is Professor Jesse Machuka of Kenyatta University with whom I set up a joint project on the development of drought-tolerant maize. Sadly he died early in May 2013. He will be sorely missed but we will ensure that his work continues as he would have wished.

List of acronyms

AATF African Agricultural Technology Foundation
AECI African Explosives and Chemical Industries
AGERI Agricultural Genetic Engineering Research Institute
ARC Agricultural Research Council
BRICs Biotechnology Regional Innovation Centres
BMS black Mexican sweetcorn
BSE bovine spongiform encephalopathy
CaMV cauliflower mosaic virus
CFT confined field trials
CMV cassava mosaic virus
CP coat protein
CSIR Council for Scientific and Industrial Research
CSIRO Commonwealth Scientific and Industrial Research
    Organisation
EFSA European Food Safety Authority
FARA Forum for Agricultural Research in Africa
FRD Foundation for Research Development
GMOs genetically modified organisms
HT herbicide tolerant
Hrap hypersensitivity response assisting protein
IAC InterAcademy Council
IDC Industrial Development Corporation
IDRC International Development Research Centre
IFOAM International Federation of Organic Agriculture Movements
IITA International Institute of Tropical Agriculture
ILRI International Livestock Research Institute
ISAAA International Service for the Acquisition of Agribiotech
    Applications
JKUAT Jomo Kenyatta University of Agriculture and Technology
KARI Kenya Agricultural Research Institute
LMCB Laboratory for Molecular and Cell Biology
MIHR Management of Intellectual Property in Health Research and
    Development
MRC Medical Research Council
MSD maize streak disease
NACI National Advisory Committee on Innovation
NARO National Agricultural Research Organisation

NASFAM National Smallholder Farmers' Association of Malawi
NBA National Biosafety Agency
NBAC National Biotechnology Advisory Committee
OECD Organisation for Economic Cooperation and Development
Pflp sweet pepper ferrodoxin-like protein
PIPRA Public Intellectual Property Resource for Agriculture
PTM potato tuber moth
Rep replication associated protein
RNAi RNA interference
SADC Southern African Development Community
SAGENE South African Committee for Genetic Experimentation
SAWISE South African Women in Science and Engineering
TIA Technical Innovation Agency
UNESCO United Nations Educational, Scientific and Cultural
      Organization
UNIDO United Nations Industrial Development Organization
UPOV Union for the Protection of New Varieties of Plants
USAID US Agency for International Development
USHEPiA University Science, Engineering and Humanities
      Partnerships in Africa
WECARD West and Central African Council for Agricultural
      Research and Development
Wits University of the Witwatersrand

Introduction

I am having lunch with Kofi Annan in his private dining room on the top floor of the United Nations. In a short while I will try to convince UN ambassadors that GM crops can help to feed hungry Africans. Kofi Annan is jovial, all smiles: 'Nothing happens till I get there.' I sweat slightly as I wait to give what could be the most important lecture I have ever given. Zambia has just said no to food aid because it might contain GM maize. Farmers might plant it, instead of eating it, even though their families are starving. But if they plant it, Zambia might lose its GM-free status for its food exports to Europe. African leaders are thus caught on the horns of a dilemma. Will I be able to convince them to embrace this technology when Europeans and others are telling them it could poison their people or, worse still, make them sterile?

· ·

From Sunday school teacher to scientist

I began teaching at the age of 13. My father ran a Methodist Sunday school in our rural Johannesburg suburb of Bryanston. There was a rough and tumble 'hall' in which the whole school would gather on a Sunday morning. Beforehand, Dad, my brother Rob and I, together with early arrivals, would collect benches from the basement of the grocery shop across the road. After assembly, during which we would sing choruses accompanied by my brother on his piano accordion, we would break up into our age groups and a 'teacher' would engage us. After a few years Dad suggested I work with one of the groups of younger children. He thought I could get rather good at this. He had noticed how the children sat enthralled as they watched me telling and acting out the Bible stories. And so began my ambition to become a school teacher, but somehow things didn't quite work out that way.

After I matriculated from high school I had no idea what to study at the University of Cape Town (UCT), but when I discovered that most of my friends were planning to do a BA degree I decided to register for a BSc, just to do something different. In my first year I enjoyed zoology the most, so I chose that as my major. In the second year the class made our zoology laboratory (the Zoo II lab) 'home', returning there whenever we had a break from lectures. We went on a field trip to Langebaan, a coastal resort up the west coast, staying in some rather dilapidated dormitories. Mornings were spent collecting specimens of coastal marine life; afternoons were taken up with identifying and classifying them. I became the champion collector of bloodworms, digging them up from the mud flats, often up to my shoulders in oozing wet sand. And in the evenings we partied!

It was during this second year that the first change in my plan to become a school teacher occurred. My favourite lecturer was George Branch. He was then a very new lecturer but subsequently rose to become a professor and head of the Zoology Department at UCT. He and his wife Margo Branch are joint authors of the widely acclaimed *The Living Shores of Southern Africa* (Branch and Branch, 1981). When George announced that he needed scorpions for a class dissection, I offered to take him into the forest above our house, where I knew these arthropods lurked in abundance. Afterwards I took him home to have tea with Mom and he asked what I planned to do after finishing my degree. 'Become a school teacher', I replied, with great certainty. He thought about this for a while and then suggested that I might consider studying further before making up my mind. During a visit to Europe at the end of that year, I visited Oxford and decided I wanted to study there. But 'No,' said George. 'Oxford is for the arts and Cambridge is for science'. 'Fine', I said, 'I'll go to Cambridge'—a life-changing decision.

My parents weren't particularly thrilled about the prospect of my studying zoology in the same department where my uncle had committed suicide some years earlier, but they bravely consented, as long as Aunt Margaret agreed. Aunt Margaret, who was really a cousin, was the headmistress of Rustenburg Girls' High School in Cape Town and, as such, had to give her approval on all academic matters in our family. When we broached the subject with her, she thought it would be a good idea. However, knowing the high standards required to gain entrance to Cambridge, she added that it was highly unlikely that I would get in. Always keen to rise to a challenge, I wrote the entrance exam during my final Zoology III exams, and was accepted.

To college at Cambridge

In September 1968, I arrived at Newnham College, the women's college at Cambridge to which I had been assigned, where I was to study for Part II of the Natural Sciences Tripos. When I first walked into the darkly imposing entrance hall of the Zoology Department, I looked at the board showing the names of staff members, and was completely overawed—it appeared just like a list of illustrious names on the cover of an undergraduate textbook.

Unfortunately, I was to find the people in the department cold and unfriendly, completely unlike the department at UCT. In addition I was assigned a rather aloof tutor, and the situation didn't improve when I received something like 40 per cent for my first essay (I was used to receiving firsts). Then I went on a field trip to Plymouth in the Easter vacation to study marine life. The weather was foul, I was seasick most of the time, and our digs were basic if not primitive. I was rapidly becoming disenchanted with zoology, a situation exacerbated by the discovery that I was expected to spend the summer vacation working in the department. What to do? My parents were coming over in July and we had planned to travel in Norway, after which I was to go on a month's sightseeing in Greece with a friend. Taking my courage in both hands, I cycled out to the Genetics Department, which was a little way out of Cambridge on the road to the cathedral town of Ely. I met the head of department, told him that I had studied genetics for about a week at UCT, and asked, 'Please can I switch from zoology?' I'll be eternally grateful to him for agreeing. And so my academic future changed dramatically.

Life in the Genetics Department was very different. For a start it taught me the importance of competition—during our lunch breaks we played 'killer' darts in winter and 'killer' croquet in summer. It also taught me a bit about the difficulties women students and staff were encountering in the male-dominated Oxbridge universities. A professor from Oxford came over once a week during the first term to teach us ecological genetics. One day a few of the male members of our class of 12 were absent, so the professor grandly told us, 'Gentlemen, we do not have a quorum; I will not lecture today', and with that he packed up his lantern slides and left! Although women were allowed to graduate from Oxford as early as 1920, this was deplored as a dangerous and radical enterprise in Cambridge. A document circulated to the Senate described Oxford's move as 'a dark and difficult adventure, the outcome of which no man can foresee'.

There was to be no female graduate of Cambridge for another 28 years (Robinson, 2009).

Even in 1968 there were only three women's colleges at Cambridge; the male-to-female ratio was about 10 to 1 (great for our social lives) and mixed colleges were not even dreamt of. I was required to eat one meal a week in college and the only reason I was happy to do so was because it gave me a free meal. I would otherwise have steered clear, as the women were somewhat frightening. They were so earnest and single-minded—true 'bluestockings'—which they doubtless had to have been to get there in the first place.

Along with a modicum of genetics, the Genetics Department also taught me the importance of speed in research. We did small projects, and initially mine focused on that workhorse of classical genetics, the fruit fly (*Drosophila*), and on fungi. *Drosophila* take about a week to grow, and although fungi are somewhat quicker they are messy, with spores that can fly around and contaminate things. I therefore figured that bacteria were the organisms to work with for really fast results: they can grow overnight from one microscopic cell to a visible colony. So I decided to do my PhD on the genetics of bacteria.

South African PhD

I returned to Cape Town facing two choices of supervisor: Prof Jack de Wet, at the University of Pretoria, or Dr David Woods, who had recently returned to Rhodes University from his post-doctoral fellowship at Oxford. Despite the fact that Dr Woods was rather an unknown entity and I had never visited Grahamstown, where Rhodes is situated, Prof de Wet was a decidedly known entity—most students regarded him as terrifying. So I went to Rhodes.

The Leather Research Institute was located in Grahamstown and they had brought a problem to Dr Woods's attention. The cured cow hides used in the production of leather were being spoiled, possibly by bacteria and their viruses. Researching these organisms and this possibility became my project. At the end of that year, I had a marvellous break. Shell SA announced that they would give a bursary to a PhD student who was working on a research project that could have national importance. I went for an interview in Cape Town, where the Shell head office is located, and I won the bursary! At the time it was the most lucrative in the country, so after a frugal existence in which I had depended on my generous Dad for support,

I now lived much more comfortably, and was even able to buy a second-hand car.

One of the advantages of doing research at Rhodes University in those days was its small size. For a period of a few months I had to add reagents to an experiment every four hours, day and night. I had no problem at all with putting on a dressing gown and a pair of slippers and going off to the lab at night to do the necessary. Years later, a former student who had shared my residence, Lillian Britten, (where I was the warden), told me she remembered this vividly, and that the whole house had thought I was slightly crazy.

Towards the end of my PhD I was awarded a Rotary Foundation post-doctoral fellowship to study in the United States. Where to go? I wrote numerous letters to a range of scientists, some quite famous and some who even bothered to reply. The problem was solved when I met Graeme Hardie, a South African living in Tucson, Arizona. During a whirlwind visit to America we became engaged. He was planning to study at the Massachusetts Institute of Technology (MIT) so I decided to try for nearby Harvard, and was accepted by the Department of Microbiology and Molecular Genetics at the Medical School.

Harvard post-doc

We arrived in Boston, newly married, in September 1974. En route we had honeymooned in Sweden and, as David Woods was on sabbatical leave in Norway at the time, he and his family joined us at a friend's home near Stockholm. He showed me a *Time* magazine article about a new technique called 'genetic engineering' in which scientists spliced genes together from any organism and transferred them into bacteria. They did this by cloning the gene into a plasmid that could be used as a vector to transfer the gene into the bacterium of choice. 'Ground-breaking stuff,' he said. Little did I know I was hearing about a technique that would become the basis for my life's work.

I had chosen Harvard mainly for its proximity to MIT and my husband, but it turned out to be an excellent choice for my career. My supervisor, Dan Fraenkel, worked in the field of central metabolism, using *Escherichia coli* (*E. coli*), the workhorse of bacterial genetics. Central metabolism, or glycolysis, is the metabolic pathway that converts sugars, such as glucose, into energy in the form of ATP. This was a totally unknown field to me. Fortunately Rick Wolf, a senior post-doctoral fellow in Dan's lab, took me under his wing and was an invaluable source of help and information. My project was to identify and map on the *E. coli* chromosome the gene coding for phosphofructokinase (Pfk), the enzyme responsible for catalysing the phosphorylation of fructose 6-phosphate to form fructose 1,6-bisphosphate. I relied heavily on Rick to understand the details of both central metabolism and the Pfk enzyme, and I would really have struggled without him.

A problem for me at Harvard was that South African PhDs follow the UK system, whereby one learns a great deal about one's thesis subject, and very little else. In America, on the other hand, they follow a far more generalised approach and students take a considerable number of intense courses, given—in the case of Harvard—by experts in the field. American PhD students are therefore well rounded and extremely knowledgeable. I felt like a total ignoramus and hardly dared to say a word in the communal tea room for at least six months. However, when I gave a seminar on my thesis, 'Genetic studies on collagenolytic *Achromobacter* strains and their bacteriophages', the topic was totally foreign to my colleagues and this, somehow, led to their respect. As a post-doc I was able to sit in on some of the courses and so managed to catch up on at least some of the body of knowledge that Harvard PhDs took for granted.

Another lesson, this time about women in science, came home to me after I had been at Harvard for more than a year. I had made friends with another post-doctoral fellow working in the lab across the corridor. We worked in different fields but found we could discuss our work easily together. It was only when she invited Graeme and me to dinner in her home that I discovered she was working under her married name. When she told me her maiden name, I realised that I had quoted her research papers a number of times in my PhD thesis. Moral: if you have published under your maiden name, stick to it. I have done so ever since.

The debate begins

The new science of genetic engineering was highly controversial. Even among scientists it was a fiercely debated topic. No-one in our department had used the technique and indeed some of them, belonging to a group called Science for the People, vehemently opposed it. The potential hazard of recombinant DNA technology had first been brought to the attention of the public by scientists—principally by Paul Berg. In their letter to the president of the National Academy of Science of the USA (Berg et al., 1974), these scientists requested that the president appoint an ad hoc committee to study the biosafety ramifications of this new technology. He did, and this committee decided that an international conference was necessary to resolve the issue and until that time scientists should halt experiments involving this new technology. Accordingly, in 1974 Paul Berg convened what came to be known as the Asilomar Conference, held at the conference centre of that name in Monterey, California. The conference concluded with a suggested set of guidelines under which rDNA (as the recombinant DNA technique became known) could be carried out, depending on the level of perceived risk. These were later fleshed out by the National Institutes of Health Recombinant DNA Committee, and became the basis of regulations worldwide.

Ironically, Paul Berg shared the 1980 Nobel Prize for Chemistry with Walter Gilbert, who received the prize for his pioneering work on DNA sequencing. Gilbert manned a stand at a weekly market with a DNA helix in hand, trying to explain recombitant DNA to passers-by. I remember his seminar on DNA sequencing, the first time many of us in the audience had even heard that it could be possible to read the nucleotide sequence of a piece of DNA—amazing stuff then, but today just a matter of course.

Although the scientists left Asilomar thinking that they had allayed public fears about their work, they had only managed to fan them. The cover of *Time* magazine on 18 April 1977 shows a sinister-looking scientist peering with obvious malicious intent at an evil-looking pink brew of DNA. In fact, if the DNA in the test tube had turned pink it would have been due to impurities in the phenol used to extract it, and the DNA would have been shattered into so many tiny fragments that it would have been unusable in an rDNA experiment. But of course a test tube containing colourless DNA wouldn't have looked half as dramatic.

The main Harvard campus is located in Cambridge, where Mayor Vellucci had a somewhat confrontational Town vs Gown approach. He used the genetic engineering controversy to proclaim that no such experiments could be carried out in his city. (Fortunately, the medical school where I worked was across the River Cam in Boston.) 'Something could crawl out of the laboratory, such as a Frankenstein,' he is reported to have said in *Time* magazine in 1977 (Jaroff et al., 1977).

The fire continued to be fuelled by scientific controversy. Caltech's Robert Sinsheimer proclaimed: 'Biologists have become, without wanting it, the custodians of great and terrible power. It is idle to pretend otherwise.' Columbia's Erwin Chargaff further built on this with 'Have we the right to counteract, irreversibly, the evolutionary wisdom of millions of years in order to satisfy the ambition and the curiosity of a few scientists?' This was countered by Harvard's Bernard Davis who was so sure the new technique was safe that he publicly offered to drink rDNA. He was supported by Stanford's Stanley Cohen, who wrote, with undisguised sarcasm, that it was Chargaff's 'evolutionary wisdom that gave us the gene combinations for bubonic plague, smallpox, yellow fever, typhoid, polio and cancer' (Jaroff et al., 1977).

A few weeks later the letters in *Time* magazine, 9 May 1977, reflected the ongoing debates (which persist to this day):

> *'Regulation seems to be necessary, if only to placate an aroused public, but the guidelines for DNA recombination should arise from within science and not from Government or other third parties.'*

> *'Once the doors to genetic engineering are thrust wide open, it will be pursued to its ultimate end: man transformed into a biological machine manipulated and controlled by the few.'*

> *'In view of the fact that our so-called human intelligence has already driven some species to extinction, there is only one commandment for us in nature's bible: Thou shalt not tamper.'*

> *'The act of living, as opposed to existing, requires the taking of chances—and perhaps a little faith. The worlds that may be opened by genetic research seem well worth a risk.'*

Illustrating the contemporary nature of these debates, the following petition appeared on the Internet in October 2011 under the headline: 'UK plant scientists call on Europe to change current laws and adopt science-based GM regulations':

> *We the undersigned share the views of 41 leading Swedish plant scientists that current legislation of GM crops is not based on science, ignores recent evidence, blocks opportunities to increase agricultural sustainability and stops the public sector and small companies from contributing solutions.*

> *We call on pressure groups and organic trade associations to cease and desist from blocking genetic solutions to crop problems, and on Europe to change current laws and adopt science-based GM regulations* (http://www.ipetitions.com/petition/changeeugmlegislation/).

The controversy is my solution

I had spent the first year of my post-doc tackling the *E. coli pfk* gene using classical bacterial genetics approaches and I was getting nowhere. But in 1976 I found the solution. I could solve the problem using the methods of that hot topic—genetic engineering. On a memorable holiday to visit the historical town of Williamsburg with Rick Wolf and his wife, I tried out the idea on him. He was very enthusiastic and encouraged me to go for it. But when I told Dan Fraenkel that I wanted to use this new rDNA technique he was somewhat sceptical (after all, he was paying) and, although he agreed, he said that I was on my own.

What I wanted to do was to clone the *pfk* gene, introduce it into an *E. coli* strain lacking this gene and show that the resulting strain now produced the Pfk enzyme. Fortunately for me, a group in the US had recently made a gene bank of random fragments of *E. coli* DNA cloned into a plasmid. I obtained this bank from a scientist in California and selected the one I wanted by transferring all the hybrid plasmids into a strain of *E. coli* lacking this gene. When I found one that could complement the *pfk*-strain and allow it to grow on glucose, I had the plasmid I wanted. I purified it and used this hybrid plasmid to prime an *in vitro* transcription/translation extract. When I tell my students today that I personally made the *in vitro* cell extract they are open-mouthed with amazement. What took me nearly six months to make and get to work, they can buy off the shelf today. However, make it I did, and then precipitated the radioactively labelled Pfk protein using a specific antibody. The moment of truth would come when I ran the products on a gel, cut this into one millimetre slices and ran the fractions through a liquid scintillation counter.

That moment of truth came exactly seven months after I had started the project. In the middle of the night on 17 September 1976, as I huddled over the scintillation counter, mesmerised by the print-out, suddenly there it was—the most perfect single peak at precisely the point at which I expected the Pfk enzyme to appear (Figure 1). Despite the lateness of the hour I just had to call Dan to tell him the good news.

The next day I wrote up the results for publication and when my first draft was complete, I handed it to Dan, with both our names as co-authors. But when he handed it back to me with the corrections, he had crossed out his name. I was mortified. I thought he had considered the paper to be so bad that he didn't want to be associated with it. When I finally plucked up the courage to ask him he replied that he couldn't claim co-authorship to the paper as it had been my work entirely, from concept, through experimental ups and downs, until the final successful completion.

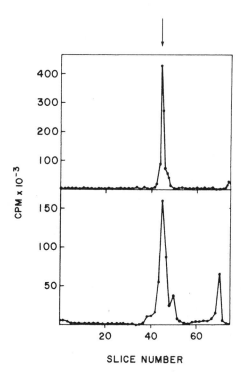

Figure I   The Pfk peak

So I am the proud possessor of a single author paper from Harvard Medical School (Thomson, 1977), an achievement reached from the starting point of a fixed desire to become a high school teacher. My path had led me from zoology to a degree in genetics from Cambridge, to a PhD in bacterial genetics at Rhodes University, and placed me neatly at Harvard Medical School just as my destined field of genetic engineering was at its start.

Frequent debates

As an adviser on genetically modified organisms (GMOs), I've had many interesting interviews over the years, and frequently had to answer the same questions. One request came in 2010 from the Canadian Broadcasting Corporation when I was at the African Agricultural Technology Foundation (AATF) in Nairobi (see Chapter 4). Could

David Suzuki interview me for a radio programme? He was a well-known geneticist, turned populariser of science through newspaper columns, television and radio, so I was thrilled. Fortunately Nancy Muchiri, head of communications for the AATF, did a background search on him. In 1977 he had written 'I feel compelled to take the position that ... no such experiments [on recombinant DNA] will be done in my lab; reports of such experiments will not acknowledge my grants; and I will not knowingly be listed as an author of a paper involving recombinant DNA'. In this personal perspective on biotechnology, put out by the David Suzuki Foundation, he further wrote: 'Feeding the starving masses through biotech in the near future is a cruel hoax that cannot be taken seriously'. And I was about to be interviewed by him on the role GM crops could play in fighting hunger.

David (we were immediately on first name terms) was very charming but clearly wasn't a neutral interviewer. He seemed to assume that the AATF was simply importing American GM crops to use in Africa. When I asked, 'David, have you ever been on a farm in Africa?', he replied that he had not, so I had to explain to him how all GM crops used in Africa have to be varieties bred for the specific climate, pests and soil conditions. At the end he said he had more hope for GM crops than he had entertained before, but I have no illusions that I changed his mind.

One subject that often comes up in debates is the issue of labelling. Some people contend that all items on sale should be labelled to indicate whether they contain any component that has been subjected to genetic modification. This means that an item will be labelled not on its content but on the way its content has been produced. If this were to occur, logically all food items should be labelled in a similar fashion. What about the pesticide sprays that have been applied to the apples in a can? What about whether the tomatoes were hand- or machine-picked? Product labelling that conveys essential information is important, but if methods of production are to be labelled, then this should be across the board. Who would monitor this? Do we even have the technological capability of testing whether what is written on the label is true? And who would pay for it?

Newspaper billboards appeared in Cape Town in March 2012 proclaiming 'GM scare in baby food'. The regulations governing the Consumer Protection Act had come into effect on 1 October 2011 and investigators had subsequently found that certain infant foods contained more than five per cent GM products, that being the cut-off level above which labelling is required. Indeed, food should be labelled (and the labelling paid for) now that South Africa has such a law, but a 'GM scare'? Considering that organisations such as the British Medical Association, the European Union Research Directorate, both French Academies (of Medicine and Science), and the Union of German Academies of Science and Humanities, among others, have declared food derived from GM plants to be safe for human consumption (including babies) (see Chapter 9), its presence can hardly constitute a 'scare'.

On the question of labelling, as early as 2009, 16 Nobel Prize Laureates warned against legislation targeting the process of developing transgenic plants, and not the product itself. They unanimously stated that:

> there is no scientific justification for additional specific legislation regulating recombinant research per se. Any rules or legislation should only apply to the safety of products according to their properties, rather than according to the methods used to generate them (Sehnal and Drobnik, 2009:13).

Another subject that often comes up in debates is the question of animal genes in GM crops. Vegetarians and vegans, in particular, are often outraged at this prospect. They cite the example where scientists introduced an anti-freeze gene from flounder fish into strawberries to try to prevent frost damage. The experiment didn't work so the project was abandoned—but not by the anti-GMO lobby. In fact, no animal gene has yet been introduced into a plant for commercial use. However, as humans share about 50 per cent of their genes with bananas, one might be hard pressed to define an animal gene.

The question of allergies is also raised on occasion. What if genes from peanuts or Brazil nuts, both foods having well-known allergenic properties, were to find their way into maize or wheat? This is taken very seriously by regulatory authorities, and applicants seeking approval of a GM crop have to provide stringent experimental and *in silico* data (many allergenic proteins contain signature amino acid sequences) to prove that their plants do not contain any introduced allergenic proteins.

A further subject that frequently arises during debates is the 'conflict' between organic farming and the use of GM crops. I write 'conflict' as I will never understand how a gene, a quintessential organic compound, cannot be compatible with organic farming. However, the International Federation of Organic Agriculture

Movements (IFOAM) states that organic agriculture dramatically reduces external inputs by refraining from the use of synthetic fertilisers and pesticides, genetically modified organisms and pharmaceuticals (IFOAM, 2005).

Interestingly enough, although organic farmers espouse the writings of Rachel Carson in *Silent Spring* (Carson, 1962), it has been pointed out that the explosive uptake by farmers around the world of GM crops, wherever they have been allowed access, is that they have brought life to the vision of the future first articulated by her when she described the new paradigm she hoped for in the relationship between humans and our environment (Giddings et al., 2012). She wrote:

> *A truly extraordinary variety of alternatives to the chemical control of insects is available. Some are already in use and have achieved brilliant success. Others are in the stage of laboratory testing. Still others are little more than ideas in the minds of imaginative scientists, waiting for the opportunity to put them to the test. All have this in common: they are* biological *solutions, based on an understanding of the living organisms they seek to control, and of the whole fabric of life to which these organisms belong. Specialists representing various areas of the vast field of biology are contributing—entomologists, pathologists, geneticists, physiologists, biochemists and ecologists— all pouring their knowledge and their creative inspirations into the formation of a new science of biotic controls* (Carson, 1962:240).

Rachel Carson pre-empted the development of insect-resistant GM crops by some 30 years!

The recent Séralini saga

On 18 October 2012, a headline in the South African *Cape Times* read: 'GM maize scare for SA after rats fed on product develop tumours'. A group in France, led by Professor Gilles-Eric Séralini, had fed rats herbicide-resistant GM maize for two years and found that they developed higher levels of cancers than rats fed on non-GM maize. Their publication in the peer-reviewed American journal *Food and Chemical Toxicology* caused a storm of controversy, with the anti-GM lobby calling for a ban on planting of such maize.

However, these voices were soon quashed by the European Food Safety Authority (EFSA), which issued a press release (http://www.efsa.europa.eu/en/press/news/121004.htm) in which it concluded that the study was so seriously flawed that the data could not be considered reliable. Similar views were expressed by six European Union member states (including France), as well as regulatory authorities in Canada, Brazil, Australia and New Zealand.

Some of the shortcomings include that:

- the strain of rat used in the two-year study is prone to developing tumours during their life expectancy of approximately two years; for this reason, the standard toxicological tests on these rats, used worldwide for products such as food additives, flavourings, etc., are done over a period of 90 days
- the authors split the rats into 10 treatment sets, but established only one control group; this resulted in some 40 per cent of the animals not having appropriate controls, a fatal flaw in any tests of this nature
- the paper did not comply with the internationally recognised standard protocols that have been developed by the OECD (Organisation for Economic Cooperation and Development)— for instance, these protocols specify the need for a minimum of 50 rats per treatment group; the authors used only 10
- the paper does not employ commonly used statistical analysis methods.

Unfortunately these comments received very little publicity compared to the original scare articles which appeared in many countries.

How could a study almost universally regarded as seriously flawed and misleading be published in a respected journal? The fact that an article is peer reviewed does not bring finality to the findings in the article, nor does peer review always meet the goals of ensuring that the results are valid and meaningful. Peer review is merely the first check in establishing the veracity of published scientific information. This is followed by analysis by the scientific community as a whole, during which the published information is more widely examined and critiqued. Finally, additional studies may be conducted when necessary to clarify any important issues raised. Clearly, the Séralini study has now gone through the second phase, and has failed this review (ABNE, 2012).

Writing on science

In 1977 I took my first step in writing on science for the public. A friend in California wrote for the American Association of Retired Persons and, when he heard me talking about genetic engineering, suggested

I write an article for the division of the National Retired Teachers Association. My article, 'Biology's "atomic bomb"', appeared in their January edition, written under my married name (Hardie, 1977). Even then I was passionate about unbundling the negative debate that was raging in America at the time, talking about the technique 'gathering its own mushroom cloud of "hot air" over itself'. I ended by saying: 'We are all passengers on earth—we come and we go; and as we pass by we, too, will leave the earth for future generations. Let us leave it a little better off than before, by the time we go.'

My second step into writing on science for the public came early in 2001. I received a call from the editor of UCT Press asking me if I would like to write a book on genetic engineering with an African slant. There were a few books on genetic engineering available for the layperson at the time. *The Thread of Life: The Story of Genes and Genetic Engineering* (Aldridge, 1996) was excellent, but went into many aspects of genetics other than genetic engineering. *Genetic Engineering—Dream or Nightmare: Turning the Tide on the Brave New World of Bad Science and Big Business* was written by Mae-Wan Ho (2000), a well known anti-GMO activist, and *Redesigning Life? The Worldwide Challenge to Genetic Engineering* was also up front in its opposition to this technology (Tokar, 2001).

In 2002 I published *Genes for Africa* (Thomson, 2002). The launch took place on the pavement outside the Village Bookshop in the main road of Plettenberg Bay, a holiday town on the southern Cape coast. I think I sold four copies.

My second book came about as a result of a review of *Genes for Africa* in the journal *Nature*. The review was written by Gordon Conway, who was then president of the Rockefeller Foundation, and in it he wrote:

> *This is a gem of a book. It is clear and concise, it makes the complex seem simple without losing the essential truths, and, as far as I can tell, it is accurate, with no innuendo, no half-truth and no wild extrapolation. [...] The remaining concerns centre on the probability of gene flow, providing wild relatives with competitive advantages that could significantly change natural ecosystems. I think the jury is still out on this, and I feel that she could have expanded further on these topics (but then I am an ecologist). [...] I recently had a long conversation with President Museveni of Uganda. He asked many thoughtful and penetrating questions about GM technologies. After our talk I sent him a copy of* Genes for Africa. *I know it will have given him many of the answers he is seeking.*

I therefore decided I needed to write another book concentrating on GM crops, considering their environmental impacts.

By this time a number of additional books had been published on the subject of genetic engineering. *Dinner at the New Gene Café: How Genetic Engineering is Changing What We Eat, How We Live, and the Global Politics of Food* (Lambrecht, 2002) talked of an impending clash between a powerful but unproved technology and gathering resistance from people worried about its safety. *Playing God? Human Genetic Engineering and the Rationalization of Public Bioethical Debate* (Evans, 2002) and *Redesigning Humans: Choosing our Genes, Changing our Future* (Stock, 2003) discussed the technology as applied to humans. *Shrinking the Cat: Genetic Engineering Before We Knew About Genes* (Hubbell, 2002) contended that the concept of genetic engineering is hardly new, as humans have been tinkering with genetics for centuries. The author focused on four specific examples: corn, silkworms, domestic cats and apples. And *The Hope, Hype, and Reality of Genetic Engineering: Remarkable Stories from Agriculture, Industry, Medicine, and the Environment* (Avise, 2004) was rather sensational. There was clearly room for a book dealing with agricultural biotechnology, written for the interested layperson.

There are two titles to my second book—an interesting result of the intervention of two publishers: Cornell University Press published it as *Seeds for the Future*, and the Commonwealth Scientific and Industrial Research Organisation (CSIRO) Press published it as *GM Crops: The Impact and the Potential* (Thomson, 2006).

This third book, *Food for Africa*, looks at the development of agricultural biotechnology in the form of genetically modified crops in sub-Saharan Africa. The *raison d'être* for the development of these crops is to alleviate hunger and achieve food security on the subcontinent.

Food security has been defined as a situation in which all people, at all times, have physical and economic access to sufficient, safe and nutritious food to meet their dietary needs and food preferences for an active healthy life (Rome Declaration, 1996). This is affected by a complexity of factors including:

- unstable social and political environments leading to war and civil strife
- trade imbalances
- natural resource constraints such as lack of water, and natural disasters such as floods and locust infestations
- poor human resources

- gender inequality
- inadequate education
- poor health
- absence of good governance.

Most of these are beyond the scope of this book, although I will touch on trade imbalances, lack of water, poor human resources, gender inequality, inadequate education and absence of good governance.

It does not take a rocket scientist to conclude that the root cause of food insecurity in developing countries is the inability of people to gain access to food due to poverty. This was highlighted in a book I co-edited for the InterAcademy Council, entitled *Realizing the Promise and Potential of African Agriculture* (2004). While the rest of the world has made significant progress towards poverty alleviation, Africa—in particular sub-Saharan Africa—continues to lag behind. Over 70 per cent of its food-insecure people live in rural areas (Heidhues et al., 2004). It follows, therefore, that a major intervention to help in achieving food security should be made in agriculture. This book sets out to show the role that the not-so-new technology of genetic modification can play in this process.

I have been fortunate to be closely involved in the development of genetic modification since 1977, first in South Africa and then in a number of mainly southern and East African countries. My involvement has been not only in the regulatory aspects of genetic engineering. I have also helped in the development of organisms, including plants. It has been, and continues to be, a long, sometimes arduous, but ultimately rewarding experience. A friend once told me that in order to be successful in developing a genetically modified crop, you don't just need a good idea (or two). What you really need is perseverance. Fortunately, the perseverance fairy was present at my birth.

Chapter 1

The SAGENE years

After my post-doc at Harvard Medical School ended in 1977, I took up a lectureship back in South Africa, in the Genetics Department at the University of the Witwatersrand (Wits) in Johannesburg. I applied to the Council for Scientific and Industrial Research (CSIR), which was, in those days, the government agency dispensing research grants, for funds to continue my work on genetically modified bacteria. When the Head of the Genetics Department, Nancy van Schaik, read my application, she wrote to the CSIR, pointing out that this was a field of research which had caused a great deal of discussion worldwide and that they might want to consider some guidelines for South Africa. She didn't know whether the people reading my application would be up to date with what was going on elsewhere. The National Institutes of Health (NIH) of the US had published their guidelines in 1976 for research involving recombinant DNA molecules and, on receiving Nancy's letter, the CSIR realised they had better set up some sort of a body to ensure that these guidelines were implemented in South Africa. They accordingly established the South African Committee for Genetic Experimentation (SAGENE). To help them in their task they invited Herb Boyer to visit.

In 1976, together with the venture capitalist, Bob Swanson, Herb had co-founded Genentech, the first biotechnology company based on GMOs, and served as vice-president of the company until his retirement in 1991. Herb, working at the University of California, San Francisco, had been one of the first scientists to discover that genes from bacteria could be combined with genes from higher organisms, eukaryotes, to create recombinant (or genetically modified) organisms. In 1977, he and his collaborators synthesised the gene coding for the human growth hormone inhibitor, somatostatin. He transferred it into the bacterium, *Escherichia coli* (*E. coli*), and developed the product in a fermenter. This was followed by the production of synthetic human insulin. Prior to that, diabetics worldwide had been treated with insulin extracted from the pancreas of animals such as pigs or cattle. Not only was this expensive, but some sufferers experienced allergic reactions to the foreign hormone. Herb's team solved these problems in one fell swoop, but helped to create a whole new problem, in the form of the soon-to-be-maligned field of genetically modified organisms (GMOs).

An extremely positive outcome of Herb's visit was the role of SAGENE in the certification of laboratories for the use of GMOs. Before a scientist could apply to the CSIR for research funding in this field, SAGENE had to approve the laboratories in question as being compliant with the US NIH guidelines. As many universities in the country were keen to foster this type of research, they were forced to upgrade and equip laboratories to a given standard. The scientists in question also had to give evidence of having been trained in the correct safety standards. This led to the running of a number of training courses, resulting in a network of scientists working on a variety of projects using GMOs. This certainly stimulated the growth of modern biotechnology in South Africa.

A visit to Basel

In 1978, a year after I had joined Wits University, I attended a life-changing three-week course on genetic engineering in the Basel laboratory of Werner Arber, who was shortly to receive the Nobel Prize for his pioneering work in this field. Werner Arber had discovered restriction enzymes which are able to cut DNA at specific sequences. This allowed other scientists to splice any piece of DNA together as long as the specific sequences at their ends matched—hence the term 'genetic engineering'.

The course was held under the auspices of the European Molecular Biology Organisation. Usually, South Africans were excluded from attending, because of our government's apartheid policy. But somehow, someone pulled strings, and against all odds I was accepted to attend the course. What an experience! Genetic engineering was only about five years old and we were learning techniques from the very people who had developed them, and using the earliest bacteriophage vectors. Among these pioneers were professors Ken and Noreen Murray. Ken's group developed the vaccine against hepatitis B, the first vaccine to be made using genetic engineering. He was also one of the founders of the UK-based biotechnology company, Biogen, and was knighted in 1993. Noreen held a personal Chair in Molecular Genetics at the University of Edinburgh and was made a Commander of the Order of the British Empire in the 2002 New Year Honours list. Sir Ken and Lady Noreen founded the Darwin Trust of Edinburgh, a charity which supports young biologists in their doctoral studies.

We were taught plant transformation by Marc van Montagu who, together with Jeff Schell and Mary-Dell Chilton, developed the first plant vectors based on bacterial plasmids and worked out how to introduce foreign genes into plants. They discovered the gene transfer mechanism between the soil bacterium, *Agrobacterium tumefaciens*, and plants, which resulted in the development of methods to convert *Agrobacterium* into an efficient delivery system for genetic engineering and thus create transgenic plants. Marc was granted the title of baron by King Baudouin of Belgium in 1990. In 1982, he and Jeff founded the biotechnology company, Plant Genetic Systems Inc., in Belgium. It is now part of Bayer CropScience. Jeff was also made a baron, and they both visited South Africa on a number of occasions to help in the development of plant biotechnology. Mary-Dell is a Distinguished Science Fellow at Syngenta Biotechnology, Inc. and in 2002 Syngenta created the Mary-Dell Chilton Center, a conference centre at their facility in Research Triangle Park, North Carolina.

We learned how to do Southern blots from their inventor, Ed Southern. These blots are used for DNA analysis and were routinely used for genetic fingerprinting and paternity testing prior to the development of microsatellite markers for this purpose. Ed also used the concept of Southern blots in the development of modern microarray slides. He founded the company Oxford Gene Technology based on this process and was made a knight bachelor in the June 2003 birthday honours. He is the founder and chair of the Scottish charity, The Kirkhouse Trust, which focuses on agricultural crop improvement research for the developing world, and specifically on legumes.

When Werner Arber's Nobel Prize was announced I wrote to congratulate him. His young daughter, Silvia, wrote a charming reply:

When I come to the laboratory of my father, I usually see some plates lying on the tables. These plates contain colonies of bacteria. These colonies remind me of a city with many inhabitants. In each bacterium there is a king. He is very long, but skinny. The king has many servants. These are thick and short, almost like balls. My father calls the king DNA, and the servants, enzymes. The king is like a book, in which everything is noted on the work to be done by the servants. For us human beings these instructions of the king are a mystery. My father has discovered a servant who serves as a pair of scissors. If a foreign king invades a bacterium, this servant can cut him in small fragments, but he does not do any harm to his own king. Clever people use the servant with the scissors to find out the secrets of the kings. To do so, they collect many servants with scissors and put them onto a king, so that the king is cut into pieces. With the resulting little pieces it is much easier to investigate the secrets. For this reason my father received the Nobel Prize for the discovery of the servant with the scissors.

One evening, walking the wet streets of Basel to dinner, Marc van Montagu asked me about my future plans. I told him that I wanted to work in the field of plant genetic engineering but didn't know how to go about it. He invited me to spend time in his lab in Ghent.

The Ledeganck Street lab, Ghent

I spent an eye-opening month with Marc's students in the Ledeganck Street lab, as we called it. Cramped would be too generous a name for the conditions they worked in. Corridors had been turned into labs, cupboards had become electron microscope units, and at times people resorted to working in shifts in order to get access to equipment. How different from Wits, where space was abundant but skilled people were in severely short supply. What I learned in that month, and from many more shorter visits to Marc's lab, set me up for my future career in plant genetic engineering, but it would take 10 more years before I finally landed in the right environment, at the University of Cape Town, to put these plans into action.

Marc was born in 1933 in Ghent in a period of great economic recession. He was raised in a working-class neighbourhood. The cotton factories where most of the men, and even young boys, worked at the time were dark, noisy and filled with clouds of dust floating around the spinning machines. As Marc wrote of these factories in his article 'It is a long way to GM agriculture': 'They were so frightening and convincingly repulsive that I felt I never wanted to be obliged to work there' (Van Montagu, 2011). Perhaps it was this that spurred him on to do well in school, helped by a school teacher uncle who insisted he go to the best primary school within walking distance. He started his PhD in the heady days of the birth of the new science called molecular genetics, but his first job was as the deputy director of an institute for training technicians and technical engineers for the nuclear industry. He soon saw the light, however, and joined the physiological chemistry laboratory of the University of Ghent. At the end of the sixties he was joined in the Ledeganck Street lab by his friend, Jeff Schell, and together with

Mary-Dell Chilton, working at Washington University, St Louis, Missouri, they founded the field of plant genetic engineering.

To get there, however, they had to find a workable plant gene vector to transfer genes into plant cells. The race was fierce between Ghent and St. Louis, with both labs using the new technique of Southern blots to demonstrate that foreign genes were integrated in the plant genome. The battle to publish first was won by Mary-Dell (we will meet her again in Chapter 7) and the only record of the Ledeganck Street lab data is a talk given at a Cold Spring Harbor Symposium in 1978 (Van Montagu, 2011).

Figure 1.1 Marc van Montagu   **Figure 1.2** Jeff Schell pointing something out on a yacht in Table Bay harbour

My own laboratory

Back in Johannesburg, all was not well with me and the Genetics Department at Wits. I was a bacterial geneticist and needed to supervise postgraduate students in that discipline. But most of the students interested in pursuing this line of research were registered in the Microbiology Department and I had no access to them. Feeling uncertain, I left in mid-1982 for a sabbatical year at the Massachusetts Institute of Technology (MIT).

This was a heady time for industrial biotechnology. New companies were starting up all around me, many involving people I was working with at MIT. For instance, Charlie Cooney, in whose lab I was working, was one of the founders of Genzyme, which began life close to the MIT lab in Cambridge and is now a multimillion dollar company (2010 revenues were in the order of US$ 4 billion) with some 10 000 employees working in countries throughout the world. They now concentrate on medical applications, but in the early days their focus—as the name implies—was on genetically engineered enzymes. Indeed, my work in Charlie's lab involved cloning the enzyme heparinase, produced by the bacterium *Flavobacterium heparinum*. Heparin is an anticoagulant used in medical procedures such as heart surgery and excessive use can cause unwanted bleeding. It was thought that heparinase, which degrades heparin, could prevent such side effects. In fact, the enzyme, whose gene was finally cloned in 1996, is now used primarily for the preparation of breakdown products of heparin for research purposes.

I got caught up in this excitement and, uneasy about my future at Wits, started a job hunt for a position with one of the start-up biotechnology companies in America. News of my enquiries reached South Africa and, before I could make any plans, I was summoned to meet Dr RR Arndt, the Deputy President of the CSIR, in Washington DC, where he was on business. He asked me whether I would like to start a Laboratory for Molecular and Cell Biology (LMCB) at Wits.

The LMCB premises started small, on the top floor of the Gatehouse Building at Wits University, in which my former home, the Genetics Department, was located. After four years it numbered some 30 people, including a group at the Onderstepoort Veterinary Research Institute in Pretoria, who were involved in animal nutrition. One of our research interests became the use of naturally occurring and genetically modified bacteria in the gut of ruminants such as cows and sheep, to improve animal nutrition.

With the memories of the start-up companies in America fresh in my mind, I decided to test the waters in South Africa for a similar venture. Together with a business friend I drew up an investment proposal for 'South Africa's first biotechnology company based on genetic engineering—AFROGEN'. The scientific advisory board was to include colleagues at Wits and Onderstepoort, and Dr Dave Woods. The projects would be largely Africa-specific and would include diagnostics for plant and animal diseases, as well as animal vaccines. The research was to be carried out mainly in the LMCB, but some would also be done in the laboratories of members of the advisory board. The estimated costs for the first three years of operation, 1985 to 1987, were R3 403 000, equivalent in today's terms to R32.8 million (about US$ 3.8 million). Start-up funding included money for market research to estimate returns on investments.

The Industrial Development Corporation and an organisation called the South African Inventions Development Corporation showed some initial interest, but after numerous meetings with business leaders, I came to realise that, in South Africa at that time, venture capital meant investment in a concept for which there were already orders in place, such as a fork-lift on the back of a truck. The suggestion was then made that the concept be modified and renamed AFROGEN Technology Transfers, which would act as an intermediary between existing research laboratories and the marketplace, closing the technology transfer gap between researchers and the marketplace. This was clearly not my field of expertise, so the idea of AFROGEN quietly died. Little did I know that in 2003 I would become the first chair of the board of a similar organisation concentrating on agricultural biotechnology, the African Agricultural Technology Foundation.

Restructuring the CSIR

During this period I reported regularly to Dr Arndt, as well as to James Bull, chief director of the National Chemistry Research Laboratory, and Brian Clark, Head of the National Institute for Materials Science. At this time, the government, which had previously paid for almost all the CSIR's expenses, decided that the organisation should become more self-sufficient, a change which Brian viewed with great enthusiasm. He played a major role in the subsequent reorganisation of the CSIR and eventually became their next president. This led to the departure of many scientists, including James and me. James became head of the Department of Chemistry and was appointed Head of the Department of Microbiology at the University of Cape Town (UCT). I wrote up this transition in 1993 with Johan Lutjeharms (previously in the National Research Institute of Oceanology at the CSIR and, by that year, Head of the Ocean Climatology Research Group in the Department of Oceanography at UCT) in an article entitled 'Commercializing the CSIR and the death of science' (Lutjeharms and Thomson,1993). I had wanted to call it 'Commercializing the CSIR and the prostitution of science', but we felt that might be a bit too provocative.

The CSIR had been established in 1945 with the express aim of bringing into existence a South African national research organisation to address national technological problems and to serve as a central scientific powerhouse. Over the years, as industries and government identified certain scientific or technological problems and recognised the CSIR's potential ability to investigate them, a number of requests for directed research were passed on to it. After 40 years, the breadth and scope of research activities of the CSIR were wide and included an eclectic mixture of basic and applied sciences. The activities of some of the CSIR's institutes were almost totally directed towards the research needs of a particular industry and could receive most of their funding from such contract research. In the ideal case, an institute would have about half its scientists at any time working on external contract work, while the other half would be occupied with more basic research, but directed towards aspects that were considered to be of national interest. Undirected basic research hardly ever exceeded 10 per cent or so of the total budget.

The CSIR thus filled an important research niche between the more esoteric basic research carried out at academic institutions and the narrowly focused product research of factory laboratories. However, by the time I joined the CSIR bureaucracy had overgrown much of the enterprise and a stultifying 'civil service' attitude of rules, regulation and forms-in-triplicate was prevalent. Indeed, that was one of the reasons I had stipulated that my LMCB was not to be established on the CSIR campus in Pretoria but at Wits University. I didn't want my staff to be run over in the 4.15 pm rush for the exit gates. So restructuring was definitely required. But what did that entail?

The 23 research institutes, some with an enviable scientific tradition, were scrapped and replaced by 11 divisions, each aimed at the perceived needs of a 'market segment'. For instance, scientists from the LMCB were earmarked to move, in groups of three or four, into about six different divisions, regardless of the need for a critical mass of skilled scientists to carry out molecular and cell biology research. As time went on, all groups as well as individuals were to be judged by financial returns or potential returns, a survival-mode way of thinking, leading to what may best be called a 'fast-buck syndrome'.

The philosophy of a 'market orientation' for science on which the changes to the CSIR were being built was not entirely original. It had previously been the vogue in Britain, the US and elsewhere. It had been widely and loudly criticised by such notables as the president of the Royal Society of London for implying that the government thought that the management of scientific creativity was no different from the management of chain stores or the running of betting shops. By the end of 1989 the pendulum had swung and the then prime minister, Margaret Thatcher, was quoted as saying: 'The greatest economic benefits of scientific research have always resulted from advances in fundamental knowledge rather than the search for specific applications' (Kenward, 1989). Present funders of biotechnology in South Africa could well take note.

It became clear that my lab was about to be abolished and the staff were told they would be accommodated in other CSIR departments— that is, except for me, for whom there was no suitable position available. We all disliked this prospect immensely as we believed that one couldn't do good molecular biology without a critical mass, which there had been at the LMCB but would certainly not be in the new dispensation. I, together with members of the LMCB, therefore approached African Explosives and Chemical Industries (AECI) for a potential solution. Why, of all saviours, approach an explosives company?

In the 1980s AECI were looking to diversify, and amino acids such as lysine came onto their horizon. Lysine is a limiting amino acid in the feed of animals such as pigs and chickens. Lysine supplementation of feed allows for the use of low-cost plant protein such as maize and soya, which are deficient in lysine, while maintaining high animal growth rates. AECI had also been supporting the LMCB financially to a fairly small but reliable extent, since its inception. The company was particularly interested in our emphasis on improved animal nutrition through natural and genetically modified ruminant bacteria. They were planning a new building to house a lysine production facility, and we suggested they build a set of molecular biology laboratories adjacent to this with a library and meeting facility in between ... and, of primary importance, they employ most of the members of our staff involved in research into ruminant digestion. To our delight they eventually agreed, and everyone (except for the powers that be at the CSIR) was extremely happy.

Thus it came about that the CSIR lost a golden opportunity to establish itself as a force in the country in biotechnology. Indeed, you will be hard pressed to find the word in their organisational structure, despite the following statement in the Department of Science and Technology's 10-year innovation plan for South Africa (2008–2018):

> *Over the next decade South Africa must work to become a world leader in biotechnology. Since the introduction of the first commercial genetically modified crops in 1995, more than 400-million hectares have been planted, 40 percent of which are grown in the developing world. And it is the developing world where the need for biotechnological innovation to solve basic problems, from health care to industrial applications, is most apparent.*

The role of SAGENE

During the LMCB years, from 1984 to 1987, the South African Committee for Genetic Experimentation fulfilled an excellent developmental role. It forced universities to improve laboratory standards in order for their academic staff to obtain research grants if they were using GMOs. They organised training courses, which had the added benefit of developing a closely knit body of scientists involved in diverse research fields but united in their use of common techniques.

In other parts of the world, especially in countries like the US and, to a lesser extent, in Europe, where biotechnology companies were springing up, regulatory authorities were constantly vigilant in ensuring that no harmful effects accrued to humans, animals or the environment. However, because GM technology was in those days mainly confined to recombinant microorganisms such as bacteria and yeast, the concerns were largely around laboratory and production plant safety. The US Recombinatory DNA (rDNA) Advisory Committee formed in 1976 was followed by other regulatory offices in the US Department of Agriculture (USDA), Environmental Protection Agency (EPA) and the Food and Drug Administration (FDA). In 1982 the Organisation for Economic Co-operation and Development (OECD) released a report for Europe into the potential hazards of releasing GMOs into the environment (Bull et al., 1982). The World Health Organization was particularly involved in safety assessment of food additives and contaminants (WHO, 1987; 1991).

As the technology improved and genetically modified organisms moved from models to commercial products, the US established a committee at the Office of Science and Technology (OSTP) to develop mechanisms to regulate the developing technology (McHughen and Smyth, 2008). In 1986 the OSTP assigned regulatory approval of genetically modified plants in the US to the USDA, FDA and EPA (US Office of Science and Technology, 1986). Countries in Africa, Asia and South America, as well as Australia and New Zealand, became seriously involved only when GM crops appeared on their doorsteps.

After some years, SAGENE felt it had accomplished much of what it was set up to achieve and therefore went into abeyance during the late 1980s, although it continued to meet from time to time. By this time I was not only a member but also the chair. However, this semi-retirement was to change radically with the advent of genetically modified crops. In 1990 we received an application from the multinational company Calgene Inc for field trials of GM cotton resistant to the herbicide bromoxynil (BXM™). These trials were permitted by SAGENE following guidelines and regulations that were applicable in the US.

Shortly thereafter, an application by Clark Cotton to conduct a US 'winter nursery production' of Bollgard® cotton seed in South Africa was also approved. Many crops, including cotton, maize and sugarcane, suffer from infestation by the larvae of certain insects. These larvae bore into the cotton boll, or the stalks, of maize and sugarcane, damaging the interiors and resulting in extremely low yields. Although some farmers use sprays to kill the pests, this is rather like shutting the door after the horse has bolted, as the spray remains on the outside of the plant while the larvae devour the insides. Moreover, as many of the cotton farmers in South Africa are poor with very small holdings, they cannot afford costly insecticides.

One of the sprays used for many years, and especially favoured by organic farmers, is based on the bacterium, *Bacillus thuringiensis* (*Bt*). This harmless soil bacterium produces a protein which is toxic to the larvae of insects. What scientists have done is to introduce the gene encoding this toxin into cotton plants. Thus, when the larvae bore into the cotton boll they ingest the toxin, which kills them. The trick is, of course, that the toxin is produced inside the plant. The basis of the toxicity is that the protein binds specifically to target cells that line the inside of the larval gut. Once bound, the protein causes the gut cells to lyse, resulting in rapid larval death. The reason that the toxin kills insect larvae only is that the guts of other animals— birds, fish and humans—do not contain the target cells to which the protein binds. The insect-resistant crops so developed are referred to as *Bt* plants, and Monsanto applied to carry out field trials of *Bt* cotton in South Africa.

Faced with these applications, the government officially reconstituted the SAGENE committee, announcing it in the *Government Gazette* of 15 May 1992. And that, followed by the road to the GMO Act in 1997, is the subject of the next chapter.

Chapter 2

From SAGENE to the GMO Act

The announcement in the *Government Gazette* of 15 May 1992, that the SAGENE committee was to be re-established, allowed SAGENE to represent the interests not only of the scientific community, but of the public as well. In addition it could advise *mero motu* (by free will) rather than only on request, and was empowered to liaise with relevant international groups concerned with biotechnology. Furthermore, it could advise on legislation or controls with respect to importation or environmental release of organisms with recombinant DNA. The terms of reference were amended on 14 January 1994 to broaden the scope from 'organisms with recombinant DNA' to 'genetically modified organisms', the term used internationally.

SAGENE was to be jointly managed by the CSIR, the Agricultural Research Council (ARC), the Medical Research Council (MRC) and the Foundation for Research Development (FRD), the agency responsible for allocating and distributing research grants. Each organisation would nominate a committee member, preferably a 'molecular biologist of international standing, active in the training and/or research in recombinant DNA and molecular genetics'. There were also representatives from a few government departments, including Health and Environment Affairs, universities, a legally qualified person to represent the interests of the public in general, a member from the SA Institute of Ecologists, and a representative of the business community or industry.

The terms of reference were that the committee would act as a national advisory body, liaise with relevant international groups, and advise on the research and application of GMOs, including possible effects on the environment as well as on their importation and/or release. The first chair was Jane Morris, whom I had recruited in 1986 to join the LMCB from her position as Research Fellow in the

Department of Microbiology at the University of British Columbia in Canada. Jane had moved to AECI when the LMCB closed and her main reason for becoming involved in SAGENE was that AECI was showing an interest in developing GM crops. In this they were following the British company, Imperial Chemicals Industry, which was a major shareholder in AECI. Jane felt there were significant regulatory gaps in the procedures for allowing the use of GM crops in South Africa and that they could prevent, or at least hinder, commercial development.

Jane was an excellent choice. Not only was she skilled in the science of recombinant DNA, but she was also a stickler for detail. It is largely thanks to her skilled leadership in those rather exploratory years that SAGENE was able to carry out its tasks so efficiently.

Establishing procedures

SAGENE's first task was to draw up procedures for assessing whether or not to approve applications for importation, trial release or general release of GMOs. I was given the job of drafting the questionnaire for this and I searched through similar documents followed by various other countries. I finally decided to base ours on the Australian questionnaire, due in part to its relative simplicity and the similarity of the Australian development of GMOs with those in South Africa. These, and the guidelines and procedures based on the UK documents, were fine-tuned by the committee and published in March 1996 as 'Guidelines and procedures for work with genetically modified organisms'. The document contained two questionnaires: one for the trial release of GMOs, which included field trials for GM crops; and one for the general release of GM plants. The risk assessment document was also based on that of the UK.

In 1994, I visited the Californian biotechnology company, Calgene, while attending a conference nearby. A few years earlier Calgene had contacted SAGENE about testing bromoxynil-resistant plants in South Africa, so I went there to discuss the SAGENE questionnaires with their regulatory team, and was delighted that they considered our regulations to be tough but fair.

GM tomatoes

Calgene had just launched the first genetically modified tomato on the market. These were the Flavr Savr tomatoes, so called as they had been genetically modified to improve their flavour by delaying fruit ripening. Calgene did this by slowing down the action of the enzyme polygalacturonase, which degrades the pectin in cell walls. This degradation results in the softening of fruit, which makes them more susceptible to damage. Unmodified tomatoes are picked before fully ripening and are then artificially ripened using ethylene gas, which acts as a plant hormone. Picking the fruit while unripe allows for easier handling and extended shelf-life, but prevents them from developing the flavour they would have if allowed to stay on the vine longer. Flavr Savr tomatoes would hopefully delay this ripening process and result in more flavoursome tomatoes. Calgene hoped that its tomatoes would taste more like the home-grown variety, picked just before eating.

I visited a local supermarket to buy some of their tomatoes. They were on a special wagon in the store, clearly labelled as having been genetically modified and every fruit had a McGregor sticker denoting the tomatoes' trade name. Unfortunately, I couldn't taste any difference between them and the unmodified tomatoes in the store, but it is my opinion that most tomatoes sold in the US taste of cardboard anyway. Sadly, Calgene was a very small company, inexperienced in the business of growing and shipping tomatoes, so this product did not last.

In the UK, the company Zeneca produced a tomato paste using similar technology. It was clearly labelled 'genetically modified tomatoes', but as it was cheaper to produce than paste made from conventional tomatoes, it out-sold the latter in Sainsbury's and Safeway supermarkets. Indeed, after about two years it held over 70 per cent of the market share, until Greenpeace initiated its vociferous opposition to GM crops by dumping a truck load of maize on the doorstep of Number 10, Downing Street. Within a very short period, sales dropped dramatically to the point where the product was removed from the market (Figure 2.1).

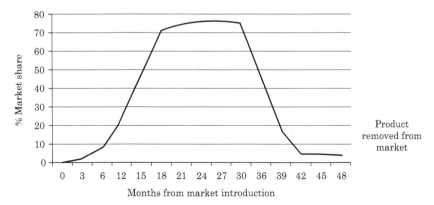

Figure 2.1   Sales of GM tomato paste in the UK
(Source: Julian Kindelerer)

I am always amazed by the double standards employed by Europe in this area. Before the onset of 'mad cow disease' (or BSE, bovine spongiform encephalopathy) the animal feed industry used 'rendered' waste animal material as high-protein ingredients in formulated animal feeds. Rendering is a process which converts unwanted animal waste at slaughter houses into useful products. Much of this is fat, but a large portion is protein for animal and pet feed. After the outbreak of BSE this was discontinued and soya became the protein supplement of choice in formulations. Where did the soya come from? Predominantly Brazil and Argentina, whose soya beans are approximately 83 per cent to nearly 100 per cent, respectively, GM (James, 2011). So although European farmers may not grow GM crops, they may happily import them and feed them to their livestock.

GM cotton

Back in South Africa, Monsanto had applied for general release of *Bt* cotton and SAGENE recommended that they investigate the value of the new varieties to smallholder and subsistence farmers. In 1997 Monsanto convinced four farmers in the Makhatini Flats region of northern KwaZulu-Natal to plant some of their *Bt* cotton seeds. At the end of that season the farmers' results were sufficiently impressive to convince more than 70 more farmers to plant *Bt* cotton. The next year, over 600 followed suit and by 2010 almost all farmers in the cotton-growing regions of KwaZulu-Natal were planting *Bt* cotton. I remember receiving an excited phone call from Muffy Koch, who ran the SAGENE secretariat. She had just inspected a field trial of *Bt* cotton and told me how, when walking through the field of non-*Bt* cotton, there wasn't an insect to be seen due to the use of pesticides, but in the *Bt* cotton field she was constantly swatting at all sorts of insects buzzing around her face. Moreover, she discovered Prinia bird nests among the *Bt* cotton plants.

From a committee to an Act

The Genetically Modified Organisms Act was published in the *Government Gazette* on 23 May 1997, and marked the beginning of the end of SAGENE. The agenda of our last meeting held on 6 February 1998 showed that applications for trials and general releases were starting to fall between two stools. While SAGENE had been made redundant by the Act, the GMO Act could not be implemented until the Regulations were approved. Indeed it took until 26 November 1999 for this to occur, and thus the GMO Act came into effect only on 1 December 1999.

Just to give an idea of the number of applications SAGENE was handling in 1997, of a total of 27 there were 13 for maize, four for cotton, two for soya, one each for canola, strawberry, eucalyptus and apple, and four for microorganisms. The applicants included companies such as Carnia, Calgene, Infruitec (local), Rhone Poulenc, Pannar (local), Monsanto, Delta and Pine Land, Novartis, AgrEvo and Pioneer Hybrid International (personal communication, Muffy Koch). The traits that were tested included insect resistance, fungal and viral resistance, and herbicide resistance. The last of these tests included resistance to glyphosate, the active ingredient in the commercially available Roundup®. Unlike many herbicides widely used in agriculture, such as atrazine, glyphosate is readily biodegradable. Conversely, when atrazine is used, maize/soya bean rotation is prevented due to the build up of residual herbicide in the soil.

Blocked by bureaucracy

SAGENE made considerable inputs to both the Act and the Regulations. We were particularly concerned that the executive council, which basically made all the decisions, consisted of members of six government departments, these being Agriculture, Science and Technology, Environment, Health, Labour, and Trade and Industry. We feared, and in hindsight correctly so, that the department members would not be well versed in the science of GMOs and, being government employees, were very difficult to pin down to attend meetings. We wanted each department to nominate experts to represent them, but this request was turned down. In retrospect, one of our initial suggestions—that the Act should fall under the Department of the Environment—which we fortunately never submitted, would have been disastrous. Our experience since the inception of the Act has been that the Department of the Environment has done all in its power to stultify the development of GM crops in the country. As it is, being administered by the Department of Agriculture presents sufficient difficulties.

Two fairly recent examples of decisions made by the Department of Agriculture will give some idea of the problems applicants face (Thomson et al., 2010). More than half a billion people around the world rely on sorghum as a dietary staple. Its tolerance to drought and heat make it an important food crop in Africa (it is indigenous to Ethiopia and Sudan). However, it lacks certain essential nutrients. In order to give it added nutritional value, the Bill and Melinda Gates Foundation funded the African Biofortified Sorghum project, run by an international consortium under the leadership of Africa Harvest, an African-based international non-profit organisation. African biofortified sorghum contains the gene for a high-lysine storage protein from barley and has increased levels of vitamin A, iron and zinc.

In 2006, scientists from the CSIR applied to the Registrar of the Directorate for Genetic Resources Management in the National Department of Agriculture, the body that administers the GMO Act, to undertake greenhouse trials of biofortified sorghum. The application was denied on the following grounds:

(1) In view of the potential risks pertaining to environmental impact (as a result of gene flow), the council recommended that this experiment be conducted on a non-indigenous species with no wild relatives in South Africa.

(2) Taking into consideration the council's concerns about gene flow, the applicant should take note that the possibility of obtaining a trial release or general release authorisation with this species—as with any other indigenous species—would be extremely low.

(3) The council expressed concerns regarding the current containment levels of the facilities that would be involved in the proposed activities and indicated that such activities should be conducted in at least a level 3 containment facility.

This decision was noted internationally. For instance, the Science and Development Network wrote on 20 July 2006 under the headline 'South Africa halts "super sorghum" study': 'South Africa has blocked trials of genetically modified sorghum that leaders of a multi-million-dollar project hope can boost nutrition in Africa'. They quote Dr Florence Wambugu, CEO of Africa Harvest, the lead institution in this multinational, US$ 18.6 million collaborative project, as saying they wished to run their greenhouse trials in South Africa because of its legal guidelines and policy framework on genetically modified crops, which are so far absent in Kenya. Interestingly, on the same day the sorghum project was put on hold in South Africa, the Kenyan parliament overwhelmingly defeated a motion by Davies Nakitare, the member of parliament for Saboti, who sought a blanket ban on all production, consumption and sale of genetically modified foods (http://www.scidev.net/en/news/south-africa-halts-super-sorghum-study.html).

In its appeal, the CSIR pointed out that the South African Biotechnology Strategy of 2002 had stressed the importance of value-addition to indigenous crops. However, the decision by the Directorate of Genetic Resources Management could be interpreted to mean that no research on indigenous crops should be allowed. The CSIR also noted that it was prejudging future applications for field trials and/ or general release and was turning down a glasshouse trial in case of a possible future application. Finally, the appeal noted that the CSIR did, indeed, have a level 3 containment facility that had been approved by that very Directorate for Genetic Resources.

Two appeals were turned down, but finally, in 2009, permission was granted. However, the damage had already been done by this slow and complicated process. The Gates Foundation moved the R&D for this project to Kenya, where approval for GM sorghum greenhouse trials was obtained within three months and trials began within five months.

The second example of a nonsensically blocked application involves potatoes. The larvae of the potato tuber moth (PTM), *Phthorimaea operculella*, bore into potato leaves, stems and tubers, causing extensive damage. In addition, fungi and mites can grow in the galleries formed by the PTM's burrowing, resulting in the decomposition of the tuber. The impact of the PTM fluctuates from season to season in response to climate, but reoccurs regularly at high levels and can cause up to R40 million (about US$ 4.6 million) in losses per annum (Visser and Schoeman, 2004).

In July 2008, an application was submitted by the South African Agricultural Research Council (ARC) for a general release of the GM potato event, SpuntaG2. This event had been developed by Michigan State University and carried the *Bt Cry1Ia1* gene (Douches et al., 2002). The required information was submitted, including socio-economic impact data and a stewardship plan. However, on 25 August 2009, the application was rejected. The reasons for this refusal included:

(1) Smallholder farmers have many other problems, and pests such as PTM might not be the most important.
(2) There is no evidence that other pest management strategies against PTM have been considered or compared with the release of GM Spunta.

(3) Entry of these GM potatoes into the formal trade is a concern. Segregation of the GM from non-GM would require an identity preservation system which is currently not in place.

(4) The capacity of small-scale farmers to implement risk-management measures could potentially be onerous.

(5) Considering the biology of potatoes, vegetative material (tubers) may be used for propagation, which may complicate risk management.

(6) Rodents, rather than PTM, are a major pest for stored potatoes.

These issues were addressed in a reply from the ARC dated 21 September 2009, appealing the decision:

(1) Information on many of the socio-economic issues can only be collected if the application is approved. This approval is needed to enable the farmer participatory evaluation, which must precede any decision on whether the ARC will use this trait for the improvement of South African potato varieties. Indeed, farmer participatory trials will help to answer many of the questions regarding the impact of the trait on potato production and farmers posed by the EC in its decision.

(2) No identity preservation system has been required for transgenic maize, cotton, soya beans, or canola used in South African formal markets for more than 10 years.

(3) The use of vegetative planting material requires no additional effort compared to the use of true seed with other crops.

(4) The levels of all potato pests vary from season to season, but PTM remains the primary storage pest.

From the above, it could be argued that the executive council had overstepped its mandate when it determined that smallholder farmers would not need this technology. It is the mandate of the ARC and farmers themselves to assess whether this GM technology is appropriate for local use. Weak decision-making processes could jeopardise the ongoing funding for this and other public sector projects. More than two years on, the appeal is still pending.

One of the unwritten reasons for both these decisions could be that the executive council was faced with a decision regarding a new crop. All previous permissions granted for field trials or commercial releases had been on GM crops already used in other countries, such as maize, cotton or soya beans. But here they were faced with nutritionally enhanced sorghum and insect-resistant potatoes, and they had no precedent of permission in other countries to fall back on. Perhaps, rather than make a mistake by allowing the confined glasshouse trial and general release to proceed, they used spurious excuses to refuse it.

In February 2012 the National Biotechnology Advisory Committee (see Chapter 5) wrote to the Minister of Science and Technology pointing out the lengthy period that appeals were taking. It recommended that the timeframes for appeals, as laid down in the Regulations of the GMO Act, should be adhered to:

> *An appeal board must be appointed within 60 days from the date of receipt of the appeal by the registrar (...) the full decision of an appeal board, together with the reasons therefore, shall be reduced to writing and furnished to the Minister, the registrar and all parties directly involved in the appeal, and made available to the public, within 30 days after the final decision has been taken.*

According to the Regulations of the GMO Act (26 February 2010), the decision-making process ought to take 120 days (90 for the appeal board, 30 for the Minister) given a best-case scenario. This means that the appeal, from start to finish, ought to take at the most 180 days. Should the Minister, however, deem it necessary, the entire process could be extended by a further 30 days, in which case the maximum allotted time would be 210 days.

As the GMO Act falls under the Ministry of Agriculture, the committee asked the Minister to bring this matter to the attention of her colleague, the honourable Minister Tina Joemat-Pettersson, as the decision-maker in the appeal process, ensuring that she understood the adverse consequences for agricultural biotechnology research, development and implementation in the country were this trend to continue. It was imperative that communication between the Registrar's office and the respective applicants was open at all times, particularly when unforeseen delays were being experienced. At the time of writing the committee has yet to hear the outcome of this letter.

Successes none the less

Despite these problems, South Africa is number eight in world plantings of GM crops (James, 2011). Most of the maize produced is consumed locally, with commercial farmers producing about

96 per cent of the crop. GM maize was introduced in 1997 but became commercially adopted on a major scale only in 2000. Plantings have increased dramatically during this time. In the 2009/10 production season, GM maize contributed 78 per cent of the total commercial area planted to maize. Of the white maize crop, 79 per cent was GM, and of the yellow crop, it amounted to 77 per cent. The average yield for the five years 1990/91 to 1994/95 was 1.9 tons per hectare, whereas that for the recent five (2005/6 to 2009/10) was 3.8 tons per hectare. Both these periods included a season of drought. Although the increase in overall yields is likely to be due to a number of factors, one of these must be the increase in GM maize planted (Agricultural Business Chamber, 2011).

Economically, soya bean is the most important legume worldwide, providing good quality vegetable protein for millions of people and animals, as well as ingredients for numerous chemical products. Towards the end of the twentieth century and into the present, soya has played an important role in helping to alleviate world hunger. Although it is currently a relatively small crop in South Africa, a survey conducted in October 2010 showed that producers intended to increase plantings by approximately 25 per cent, from 311 450 hectares to 390 000 hectares for the 2010/11 production season. Should these intentions be realised, it will be the largest area in South Africa planted to soya beans on record. Approximately 85 per cent of soya beans produced in South Africa are GM. This began to be commercially planted in 2000 and, although the differences are not as dramatic as with maize, the yields in this period increased from 1.2 tons per hectare to 1.7 tons per hectare (Agricultural Business Chamber 2011).

Small-scale farmers in KwaZulu-Natal were quick to take up *Bt* cotton. Following the success of the first four farmers who planted this in 1997, 75 took part in 1998. The uptake in 1999 was 411, 644 in 2000, and by 2002, more than 2000. However, in 2007/8 the nearby cotton gin closed down, forcing the farmers to transport their crop about 600 kilometres to the closest gin. In addition, the area under cultivation is not ideal for the growth of cotton and the crop became internationally uncompetitive, even with the addition of the insect-resistance trait. Understandably, fewer farmers are now growing cotton in that region, with only about 300 involved in the 2009/10 production year (http://www.cottonsa.org.za).

The anti-GMO lobbies, such as Biowatch, were by no means silent during this period. A group of them brought to the country a Canadian farmer, Percy Schmeiser, to speak about his experience at the hands of Monsanto. He addressed a special parliamentary hearing held in Cape Town, to which I was invited. Mr Schmeiser told us how in 1997 he discovered that a section of one of his canola fields in Saskatchewan contained canola that was resistant to the herbicide Roundup. Monsanto had been marketing GM canola resistant to this brand of herbicide under the name of Roundup Ready. Farmers were growing this variety to control weed competition. However, Mr Schmeiser, who was vehemently opposed to this technology, had not bought any of this seed. The following year, over 95 per cent of his crop of approximately 400 hectares (1 000 acres) was identified as being of the Roundup Ready variety. Monsanto sued Mr Schmeiser for patent infringement, by failing to obtain a licence for their canola seeds. Mr Schmeiser's argument was that Monsanto's seed must have escaped from passing trucks or arisen from accidental pollination. Moreover, this seed had now 'contaminated' his own strains of canola, which had taken him 50 years to develop.

During question time at the parliamentary hearing, the chair gave the floor to Andries Botha, a maize farmer from the Free State and at the time a member of parliament for the Democratic Alliance and the shadow Minister of Agriculture. Mr Botha expressed amazement that one variety of canola, Roundup Ready, could spread so rapidly across a 400-hectare field, while Mr Schmeiser was at the same time able to maintain the integrity of his own varieties of canola on other parts of his farm. How was this possible? Unfortunately the Canadian was unable to answer this simple question from one farmer to another. Also, unfortunately for Mr Schmeiser, Canada's Supreme Court ruled in favour of Monsanto.

The International Service for the Acquisition of Agribiotech Applications

An organisation that is very helpful in providing statistics on the spread and uptake of GM crops worldwide is the International Service for the Acquisition of Agribiotech Applications (ISAAA). In 2001 the president, Clive James, asked me to become a member of his board. This exposed me to the state of agricultural biotechnology in developing countries other than on the African continent.

For a workshop ISAAA held in Bangkok on intellectual property related to GM crops, I was asked to prepare two presentations entitled 'How to keep a laboratory notebook' and 'How to monitor a genetically modified crop during trials and after general release'.

Some years later, the work I put into that workshop paid off when I was involved in editing a book on intellectual property. It was spearheaded by Anatole Krattiger, who had organised the Bangkok workshop, and was jointly organised by the Oxford-based Centre for the Management of Intellectual Property in Health Research and Development (MIHR) and the University of California-based Public Intellectual Property Resource for Agriculture (PIPRA). I remember walking with the group of editors to the MIHR offices in Oxford when Anatole took me aside and asked me if I'd consider chairing the meeting. I was amazed, as of the group assembled I probably knew the least about the subject. Fortunately, Robert Mahoney, who had played a lead role in the establishment of the MIHR, and who oversaw the development and implementation of IP management policies for the Ford Foundation, co-chaired with me and in 2007 the IP handbook was duly published. It runs to a mere 1998 pages in two volumes (Krattiger et al., 2007). I will return to the issue of IP in Chapter 4 as it has become a highly contentious issue in the development of GM crops, particularly in Africa.

The 2010 ISAAA board meeting was held at Los Banŏs in the Philippines, where the head office is situated. During our time there we visited field trials of GM insect-resistant, *Bt* aubergines (also known as egg plant and brinjal). We went through a 'Fort Knox' type clearance to be allowed access and had to put on special boots before entering the fenced-off enclosure. We were allowed to walk only along the outside perimeter of the fields but were able to see the workers at their tables. They would harvest every ripe aubergine from both the *Bt* and the control, non-GM plots. Each fruit was weighed, cut in half, the insects extracted, measured, their instar noted and then all were discarded into separate metal containers: one for the *Bt* fruit and one for the controls. We arrived fairly late in the afternoon so this process had been going on all day. You would not have liked to go near the container with the control aubergines, they were rotting so badly. The *Bt* brinjals had some insect damage but it was minor. Apparently most aubergine farmers, certainly in that region, spray their crops every second day with highly toxic insecticides. And to think that the Indian government had a few months earlier denied applications for the release of *Bt* aubergines. When one thinks of the health impact on the sprayers (let alone the lost income to the farmers), that decision has to be little short of criminal.

Interestingly, a few days earlier a van-load of visitors had approached the guards at the entrance to the site, requesting permission to enter. They were not prepared to give their names or affiliations, so they were denied access. One wonders if they had harmful intent—elsewhere in the Philippines, trials had been vandalised.

At the time of this visit, Brazil had a completely different approach to GM crops. I had attended a meeting in Rio de Janeiro in about 2001, organised by the Brazilian Department of Agriculture. Its aim had been to come up with a set of recommendations for their government as they debated the growing and use of GM crops. How different their decision was. Brazil is now the world's number two planter of GM crops, just behind the US, specifically in herbicide-tolerant soya beans, of which they are a major exporter.

ISAAA's main claim to fame is as an information agency. It produces an annual brief entitled 'Global status of commercialized biotech/GM crops', which is somewhat of a gold standard on the subject. There are chapters on all countries growing significant crops of GM plants, describing what is grown, how the numbers are increasing, and so on. In the section on South Africa there is a detailed account of the established GM crops (maize, soya bean and cotton) by trait (insect resistance, herbicide tolerance and stacked traits) and an account of approved commercial releases, field trials and greenhouse trials. In addition, it relates farmer testimonies, such as from Samuel Moloi, who grows 63 hectares (156 acres) of maize on land that he rents in the Free State province. He plants GM seeds that are both insect resistant and tolerant to the Roundup herbicide. He says he spends less on diesel by using his tractor less, and less on labour, because he doesn't have to hire workers to cut the weeds. 'The GM seed is a little higher (in cost), but it does a fantastic, a wonderful job for me,' he said. 'The benefits at the end of the day outweigh the cost of the seed itself' (James, 2011).

These benefits were echoed in a recent study by Gouse (2012), who studied smallholder GM maize farmers in KwaZulu-Natal over a period of eight seasons (2001/02 to 2009/10). The results showed that *Bt* adopters enjoyed higher yields than their conventional maize-planting counterparts and, in most seasons, were better off despite paying more for their seed. In the case of herbicide-tolerant (HT) maize, farmers also benefited through higher yields. Interestingly enough, 'Farmers seem to be willing to pay for the weed control convenience of HT maize and, based on adoption figures, farmers value the yield-increasing and labour-saving benefits of HT maize higher than the borer control insurance of *Bt* maize'.

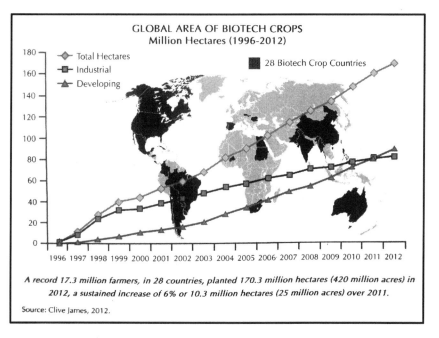

Figure 2.2   Global area of biotech crops 1996–2012

The ISAAA also publishes the Pocket K series on subjects such as *Bt* wheat, *Bt* rice, marker-free GM plants, and briefs on specialised topics, such as the 2009 one, 'Communicating crop biotechnology: stories from stakeholders', which includes some fascinating accounts from farmers who grow GM crops in various parts of the world (www. isaaa.org). I related one of these stories in Rome when I addressed the Pontifical Council for Justice and Peace. I had been warned that the Italian Minister of Agriculture was opposed to GM crops. He swept in with his entourage some time during our deliberations and proceeded to inform the audience that it had been proven that GM crops could not possibly help smallholder farmers. I spoke after him and said that it was a pity he had not arrived an hour earlier as he would then have heard a smallholder farmer from South Africa wax eloquent on how GM cotton was helping her to improve her productivity. She had told the audience that, being a school teacher, she could only spray her crops on a Saturday, Sundays being reserved for church attendance, and if she missed a Saturday by taking a child to the clinic or some other task, she knew she would lose a portion of her crop to insects. Now, however, she didn't have to spray, and 'Look at my hands', she had said, '... they don't even look like a farmer's hands any more!' The minister was not impressed.

AfricaBio

South Africa has its own version of ISAAA in the form of AfricaBio (www.africabio.org). In 1999 I received a call from a former PhD student, Jocelyn Webster (also a member of the LMCB), to say she wanted to start an organisation aimed at educating government officials, regulatory authorities, the media and the public at large about agricultural biotechnology. Jocelyn had been a member of SAGENE and was closely involved in writing our position paper on the GMO Act. Her immediate target was South Africans but in time she aimed to reach out to other African countries. I joined the organisation, together with other academics, farmer organisations, grain traders, biotechnology companies, seed companies, food manufacturers and retailers, and consumers. AfricaBio was officially registered as a non-profit, Section 21 company in 2000.

Over the years AfricaBio has proven its worth as a provider of accurate and objective information on biotechnology to consumers, media and decision-makers. It has provided a regular forum for exchange of information not only between South Africans but between people from many SADC countries. Its workshops, to provide information and training to stakeholders from countries such as Malawi, Namibia, Zimbabwe and Mozambique, have been particularly successful. It has also run training and advice programmes for small-scale farmers interested in planting GM crops.

In 2002 AfricaBio was awarded South Africa's prestigious National Science and Technology Forum (NSTF) award for its outstanding contribution to science and technology. In its citation, the NSTF stated that the organisation had provided a forum for informed debate on biotechnology issues and the promotion of its safe, responsible and ethical use, with significant contributions in the areas of education and the public understanding of science, engineering and technology.

Over the years AfricaBio has put out a series of position papers on such issues as GM and biodiversity; the impact of GM on biodiversity; bio-ethics; intellectual property rights and farmer's rights; and GM impacts on sustainable agriculture. It later produced booklets such as *Agricultural Biotechnology: Facts for Decision-Makers* and *Biotechnology: Biosafety, Food Safety and Food Aid*. For many years it has sent out the monthly newsletters called *BioLines*, followed by

GMO Indaba and more recently *GMO Insight*, which are quick guides to what is topical at the time. A few examples of its articles are listed below:

- The impact of biotechnology on Africa in the 21st century (June 2001—a meeting held in preparation for the World Summit to be held in Johannesburg in September 2002)
- China surges ahead of India in Biotech race (February 2002—in terms of research and development, not yet commercialisation)
- Zambia launches its first biotech outreach society (July 2003—and they're still working on it)
- SA GMO maize crops set to grow (April 2004—and they are still growing)
- International pressure group Greenpeace warns Philippine authorities that biotechnology 'can lead to millions of dead bodies, sick children, cancer clusters and deformities' (April 2004—and still the misinformation keeps coming)
- Tanzania jumps on GM bandwagon—Agricultural Ministry says they cannot afford to be left behind (March 2005—but it seems they are)
- Golden rice provides increased Vitamin A (March 2005—but needy children in Asia are still waiting for it)
- Kenyan minister asks journalists to highlight biotech benefits (June 2006—and some of them got it right)
- UK farmers optimistic about GM crops (February 2008—unfortunately their politicians think otherwise)
- *Bt* toxin resistance: an evolutionary action (March 2008—a cautionary note on responsible stewardship of the new technology)
- *Bt* awareness campaign for Kenya launched plus Kenya approves GMO bill (April 2009—Kenya making great strides forward)
- Consumer Protection Regulation effective October 2011 (October 2011—all food in South Africa containing more than five per cent GMO ingredients to be labelled)
- AfricaBio and partners host successful IRM workshop (December 2011—ways to prevent insects from developing resistance to the *Bt* toxin)

Although I am no longer on the board of AfricaBio, I am proud to have been part of this impressive organisation.

AfricaBio has also organised a number of debates on the subject of GM crops. As luck would have it, an occasion arose when Dr Florence

Wambugu, CEO of Africa Harvest, and Vandana Shiva, the Indian anti-GMO activist (see Chapter 4), were both in Johannesburg at the same time. AfricaBio took the opportunity to arrange a public debate in which each was joined by three like-minded colleagues. I was on Florence's team. The chair, a well-known TV presenter, explained the ground rules. After opening statements the debate would be open to the floor. The side to whom the question had been posed would be given two minutes to reply, thereafter the other side could comment. Florence's daughter was in the audience and had the following question for Vandana Shiva's side: 'The pro-GM side has said what they plan to do to improve crop production in the next five years. What does the anti-GM side plan to do in the same period?' Ms Shiva's side spoke to the question but after the allotted two minutes the TV presenter said: 'It's obvious that you cannot answer the question so it's over to the other side to respond.' At times like this one is grateful for a good chair.

Chapter 3

Into Africa

My first move into Africa north of South Africa's borders came about shortly after our return to South Africa from Boston. In 1977 my husband moved to Mochudi, a small village just north of Gaborone, the capital of Botswana. He had a Fulbright Fellowship to do his PhD there on various aspects of low-cost housing.

Playing in another field

It was not long before I became embroiled in a local scientific controversy, although this time it had nothing to do with genetic engineering. I noticed that Graeme's many mosquito bites often took rather longer to heal than would be expected, and, talking to neighbours in Mochudi, discovered that many of them suffered from other types of wounds which did not heal easily. Being a microbiologist, I wondered if the water was infected with faecal bacteria, such as *E. coli*, the signature bacterium for sewage contamination of drinking water. I took some samples back to the lab and, indeed, found some indications of *E. coli*. I contacted the Botswana Department of Health but they showed no interest, so I discussed it with friends who worked for the local hospital, which treated many TB patients. Their sewage sometimes overflowed into a small stream that ran past the hospital and they expressed concern about my preliminary findings. I therefore brought a very willing team of postgraduate students from Wits University to Botswana to do some sampling in the area.

Their first hurdle came at the border. I had forgotten to tell them that Botswana had just banned the import of alcohol across the border from South Africa. Being normal students, they had a fair amount with them, and, confronted with the problem, proceeded to drink most of it there and then. Again, being normal students, they were also impecunious, so what to do about liquid refreshments once they were in Botswana? Ever resourceful, one of their number, an experienced gambler, solved the problem after one night's hard work at the local casino.

Another hurdle, which I had fortunately anticipated, was how to incubate the Petri dishes after we had spread our water samples on them. Faecal bacteria grow at body temperature, around 37°C, a far cry from the ambient temperature overnight in a Mochudi winter. Fortunately, a friend who worked at the local museum allowed us to set up a temporary dormitory and 'laboratory' in one of his spare rooms. We rotated shifts during which each person donated a sleeping bag for a few hours, which was draped around a box carrying a few lit candles. Rather Heath Robinson, but it worked.

Back in a proper laboratory, we confirmed an unacceptably high level of faecal bacterial contamination in some of the water samples. With the spirited support of some of the local women, I called a meeting in the museum to share our results. Imagine our surprise when the local chief arrived. And that set the cat among the pigeons as word of this reached the ears of the government. Questions were asked in parliament and soon I had the Ministry of Health asking me to conduct a series of tests in various towns around Botswana. Moreover, they would supply me with a mobile lab, a driver and a laboratory assistant!

As a result, during the July vacation in 1978 my small band visited towns such as Malepolole, Serowe, Mahalapye and Palapye. The problem in most cases was that the building regulations that demanded certain distances between human habitation and drinking water supplies had not been adhered to. The authorities promised to prevent this in future and install chlorination plants, but whether this actually happened or how long it lasted I never did discover (Figure 3.1). Interestingly enough, the water below Mochudi's hospital was relatively clean as the river was very shallow and ran over rocky surfaces under Botswana's brilliant sunshine. As any microbiologist will tell you, this is an excellent natural sewage treatment system.

Pariahs stick together

The 80s were extremely barren years for South African scientists wanting contact with colleagues in other African countries, and in those early years of biotechnology, pariah states like ourselves, Taiwan and Israel were forced to stick together. It would appear

She paved the way for safe borehole water

A Johannesburg woman's concern about her husband's upset tummy has resulted in her carrying out scientific tests on the borehole water of seven Botswana villages.

Dr Jennifer Thomson, a microbiologist at the University of the Witwatersrand, recently returned from Botswana with the news that that country's Department of Water Affairs is to act on her findings and chlorinate the borehole water in the villages.

Suspicion

"My husband Graeme Hardie is doing his doctoral research in social anthropology (through Boston University in the United States) in the pretty little village of Mochudi, about 45 km from the capital of Gaborone," said Dr Thomson.

"I visit him as often as I can and was concerned when one day he developed gastro-enteritis. I sampled the water and

The Botswana Government is to chlorinate the borehole water in many of its villages as a result of tests carried out there by a Johannesburg woman, Dr Jennifer Thomson. SUE GARBETT reports.

brought it back to my laboratory at Wits for investigation."

The results confirmed her suspicion that the water was contaminated with bacteria.

"We suspected the source of the contamination could have been the hospital because untreated sewage from it was entering a dam, seepage from which could have been getting into the underground water supply feeding the boreholes," said Dr Thomson.

She chatted about the results of her casual investigation to her students, who suggested they all go on a field trip to Mochudi.

"We set up a lab in the village museum and the results of our investigations confirmed my earlier tests.

"Naturally the Department of Water Affairs was vitally interested in my results and asked me if I would return to do a fuller investigation of other village water supplies."

The Department of Water Affairs had been conducting tests for bacterial contamination based on World Health Organisation guidelines which had indicated an absence of human fecal contamination.

However, Dr Thomson showed that Botswana was unusual in that the organism that indicates the presence of fecal contamination (E Coli) is largely absent from borehole water, but that the harmful organisms were present.

"This is a most unusual,

although not totally unprecedented situation. It has occurred in Kenya and Israel," said Dr Thomson.

So for 10 days last month she travelled hundreds of kilometres in a Land-Rover with a mobile laboratory through Botswana, testing water.

She worked with a chemist from the Geological Survey Department and two technicians.

"We often worked from 5 am to 7 pm and must have covered about 50 boreholes.

Teasing

The findings have revealed that on the whole the water is pretty bad in all the boreholes as a result of which the Department of Water Affairs is already chlorinating the water.

"My husband no longer gets upset tummies. But he who is supposed to be doing research in Botswana and yet it's mine which has inadvertently turned out to be beneficial," said Dr Thomson.

4 Aug. 1978

Figure 3.1 Article appearing in *The Star* newspaper

that Taiwan at least was feeling as isolated as South Africa. In the preface to the proceedings of the Republic of China-Japan Symposium on Biotechnology, which I attended in 1987, one of the organisers wrote: 'In 1982, the Republic of China decided that biotechnology should be one of the eight major thrust areas in scientific research and technology development of this country.' He went on to say that this symposium was one of the efforts in reaching this goal, and that they were able to invite eight distinguished scientists from Japan. In addition they were 'happy to see Dr Jennifer Thomson from the Republic of South Africa attend the symposium'.

Similarly with outcast Israel. In 1986 a delegation of South African scientists held a joint meeting at the Hebrew University in Jerusalem. The timing was superb as it coincided with the 70th birthday of Ephraim Katzir, fourth president of Israel and a founder of the

Weizmann Institute of Science, named after Chaim Weizmann, who, apart from being the first president of Israel, was a biotechnologist of note. During the Second World War, while working in Manchester, he developed the Weizmann Process whereby the bacterium *Clostridium acetobutylicum* produces acetone and butanol, much sought-after in the manufacture of explosives. It was fitting that Dave Woods, himself a world leader in the genetics of *C. acetobutylicum*, was a member of our team.

Shortly after this Israeli experience, the LMCB (discussed in Chapter 1) was disbanded. This left me without a job. But most conveniently, Dave Woods, my former PhD supervisor, became Deputy Vice-Chancellor of the University of Cape Town, leaving his position as Head of the Department of Microbiology vacant. I applied. The other three applicants were men and current members of the department. Dave's parting words to me as I left his office for my interview were 'Don't screw it up'. I didn't, and on 1 April 1988 I took up my new position in Cape Town, back home after an 18-year absence.

The potential of African students

A watershed came for South Africa on 2 February 1990, when the apartheid government announced that it was unbanning the ANC and other political organisations, and would release Nelson Mandela. I was visiting one of Cape Town's black townships on that day, as I was involved in opposition politics, and well remember the extreme rejoicing in the streets. One of the results of this decision for me was that I was invited to act first as external examiner for the University of Zimbabwe's MSc in Biotechnology, and then for the University of Botswana's BSc in Biology. The former had a life-changing effect on me and the latter opened my eyes to the harmful effects that government policies on tertiary education can have on the quality of that education.

The University of Zimbabwe's Class of 1992 consisted of eight students, including three who became successful PhD students in my lab. The first was Tichaona Mangwende, who played a vital role in the early days of our work on developing maize resistant to the African endemic maize streak virus (see Chapter 7). The second was Thabane Dube, who, to our great sorrow, died of hepatitis B the day after he graduated with his PhD. The third was Dahlia Garwe, who pioneered our work on maize tolerant to drought (see Chapter 8). They were followed by two further Zimbabweans, but only Dahlia, who is currently the acting managing director of the Tobacco Research Board in Harare, has returned to her homeland. The others found nothing there to entice them back and are all working in South Africa, a great loss to Zimbabwe's scientific community.

All my Zimbabwean postgraduate students were exceptionally talented and this led to a rather interesting encounter. In the early 2000s I was working on a report entitled 'Realising the promise and potential of African agriculture' commissioned by the InterAcademy Council (IAC), a council of national science academies. Members of the Southern African Development Community (SADC) involved in agriculture were meeting in Pretoria to discuss what should be included in this investigation. Each sat behind the name of his or her country, and at the first tea break I approached the woman sitting at the Zimbabwe table. I extolled the scientific excellence of my three PhD students from her country, saying what an excellent secondary education they must have experienced there and that I hoped Mugabe wouldn't 'stuff it up'. She was his sister.

Another outcome of my visit to Zimbabwe was an invitation from the university there to run a practical course in biotechnology, funded by the United Nations Industrial Development Organization (UNIDO), during December 1993. About 20 people from a number of African countries attended and we tried to expose them to all the basic techniques available at the time. This was an eye-opener for me into the intense thirst for knowledge and extraordinary capacity for hard work shown by students when they are really driven to learn.

Why was my Zimbabwean experience so life-changing? Word began to spread through the extremely small biotechnology community in the region that Jennifer Thomson's laboratory was a good place to gain experience. Before long I began to attract students from Mauritius, Botswana, Kenya and Uganda, and with the students came exposure to their home institutions, their supervisors and their colleagues. Other organisations helped as well, but that initial exposure in Harare was critical.

My experience in Botswana was not as uplifting. The government had just decreed that the size of its tertiary education intake would be dramatically increased. By the second and third year of my assignment I began to see the effects of this. Lecturers simply could not cope, classes became impossible to handle in the space available and, inevitably, standards dropped.

During 1993 I also served on the African Academy of Science's advisory panel for their programme called 'Educating girls and women in Africa in science'. Ten of us from nine African countries (I was the only white woman) met in Nairobi, our chair being Maki Mandela, Nelson Mandela's eldest daughter by his first wife. It was fascinating, at such an early time in South Africa's democracy, to be working with women who were vice-chancellors and ministers of education. We hoped that the research projects we put in place throughout the continent would come up with results that could help to change policy in order to enable more girls and women to study and embark on careers in science. Apart from the personal achievements of successful women scientists, economic development is intimately linked to scientific and technological development, and to have almost half the population often denied access to careers in this field makes no sense at all. This was probably my real awakening to the plight of women scientists in Africa.

South African Women in Science and Engineering

With this experience behind me, I was of course receptive to the suggestion by a friend at UCT, Lesley Shackleton, that we form South African Women in Science and Engineering (SAWISE). As Lesley described her reasoning behind this initiative:

> *Not for the first time we found ourselves, a group of women scientists in the Western Cape, talking together at a research-related social event about the lot of women in the professional work place. Why were so few women heads of departments? Why did so few sit on the major funding agency, the Foundation for Research Development (FRD) and other committees that held power over our research careers? Why were so few of the oral presentations at scientific conferences made by women? And why were we doing nothing about it?* (Shackleton, 1997:12).

An initial meeting, held at UCT in February 1995, was attended by some 50 women who gave a clear mandate to an interim committee to establish SAWISE. I was elected the first chair of the Western Cape branch and a year later the Gauteng branch was formed.

The aim of SAWISE is to strengthen the role of women in science and engineering by:

- raising the profile of women scientists and engineers
- highlighting and addressing the problems faced specifically by women in these fields
- lobbying for the advancement of women in science and engineering

- providing leadership and role models for young people wishing to enter the fields of science and engineering.

Again, in Lesley's words:

SAWISE was acutely aware of the importance of science and technology for development, and the role women can play. Women have the greatest influence on the next generation, they set the standards for health and hygiene, they comprise the majority of the agricultural labour force. Taking science subjects should be seen not only as a means to a vocation, but as a means to build up the scientific and technological culture necessary for development, and a way of empowering people to apply basic scientific concepts in their everyday lives (Shackleton, 1997:12).

In those early days the projects we undertook were those that required hard work and enthusiasm, but no funds, as we had none. Members ran 'Nights at the observatory' to show school girls the wonders of astronomy; courses on 'Gender and professional identity'; and workshops for young trainee science teachers to raise their awareness of gender in the teaching of science. We were also given a donation to award annual Honours bursaries—the Angus SAWISE scholarships—and many of the recipients are well on their way to successful careers in science and engineering.

Landing in Kenya

Another direct result of the unbanning of the ANC was my introduction to Kenya. It was announced that, as the ANC had been unbanned and Mandela was to be released, South African Airways (SAA) could start flying to Kenya. I was on the very first SAA flight that was allowed to land in Nairobi, where I had been invited to give a series of lectures.

That experience taught me one very important lesson—never arrive in a new country on a weekend if you are expecting your visa to 'be waiting for you on arrival'. It wasn't, and it being a weekend, no-one was at work at the University of Nairobi (whose Professor of Veterinary Pathology and Microbiology had invited me to lecture), and therefore no-one knew anything about me. To make matters worse, it was the wedding day of the Attorney General's daughter, and everyone who was anyone, including the Vice-Chancellor, was out at the farm. Thus it was decreed by the Department of Immigration that I should be put on the first plane back to South Africa.

Somehow the news of my predicament filtered through to my extremely anxious friends waiting for me in the arrivals hall. They started to move heaven and earth to get me released, at least into their custody, until Monday (it now being Saturday) when they hoped to be able to reach the university authorities. No such luck, but at least at some time in the small hours of Sunday morning I got a message that one of the deputy vice-chancellors had come to the airport to try to help, but to no avail.

In the meantime I had settled down to make the best I could out of the situation. And the best soon arrived in the form of two Ugandan brothers who had just been to visit their aging mother for the first time since being driven out of the country by Idi Amin. They were both living in Canada and one was a veterinary microbiologist in Saskatchewan. On hearing my sorry tale they took pity on me, plied me with food and drink (I had no Kenyan currency with me) and regaled me with stories of their persecution in, and subsequent flight from, Uganda. In due course, their flight departed. By now the barman had heard my story so he kindly settled me in a corner of the bar furthest away from customers, covered me with a blanket and promised to try and keep the noisiest of the travellers away.

Eventually I was released on the Sunday afternoon. Negotiations between the Vice-Chancellor and the Department of Immigration were successful and I was free to embark on my lecture series.

The University Science, Humanities and Engineering Partnerships in Africa

Another result of the 1990 unbanning of the ANC was that the University of Cape Town in general began to be looked upon by academics in the rest of Africa as a potential university to send PhD students to. Up until then, promising graduate students were often sent to do their PhDs at universities in Europe or the US. The chances of their returning to their countries of origin after these experiences were slim. Would UCT be different? Dave Woods, as Deputy Vice-Chancellor for Research, took up the challenge and, together with Lesley Shackleton, my co-founder of SAWISE and project director for international students, visited a number of universities in sub-Saharan Africa to invite their PhD students to UCT. The upshot was a most successful programme, the University Science, Humanities and Engineering Partnerships in Africa (USHEPiA).

One of the criteria for acceptance into the programme is the guarantee of a job at the home university, so the return rate is excellent and a brain drain is alleviated. By the end of 2012, 53 PhD fellows had graduated from home universities in Kenya, Uganda, Tanzania, Zambia, Botswana and Zimbabwe. Another strength of the programme is that supervisors from UCT are required to travel to meet their prospective students, together with their home supervisors, at their home institutions, thus forming strong linkages. Accordingly, I capitalised on my contacts in Botswana and together with a colleague in my department enrolled our first fellow. The programme aims to have the student spend about half of his/her time at UCT and the rest in the home institution. Unfortunately, we naively embarked on a molecular biology project of a rather complicated nature—not the wisest choice for the fledgling programme—and special dispensation had to be made for this first student to do virtually all his lab work at UCT. However, we learned from the experience and I carefully tailored the next USHEPiA projects to allow for adequate time outside UCT.

In 2001, I was asked by the USHEPiA office to spearhead an initiative to expand the programme from the current PhD exchange to shared research projects in agricultural biotechnology. Over a period of a few days at the Jomo Kenyatta University of Agriculture and Technology (JKUAT) in Nairobi, colleagues and I put together the outlines of a number of joint projects in Kenya. We then went on to the University of Dar es Salaam in Tanzania, the University of Zambia and Makerere University in Uganda. As a result of these visits, Viviene Matiru from Kenya and Betty Owor from Uganda completed PhDs in my laboratory.

Viviene is a senior lecturer in the Department of Microbiology at JKUAT, and until recently was its chair. Microbiology is actually a sub-department of the Botany Department and principally teaches medical microbiology students. I can hear her frustration, which reminds me of mine when I was in the Genetics Department at Wits University, when she tells me that her aim is to establish a fully fledged Department of Microbiology and Molecular Biology, to be—as she says—'a centre of excellence in microbiology, to be the place of choice to study microbiology in Kenya and in the eastern and central African regions'.

Betty Owor went to Makerere University in Kampala intending to study medicine, but, fortunately for us, landed up in agriculture instead. That education stood her in excellent stead for her PhD work as it involved field trips to smallholder farmers, commercial farmers and factories. At one stage we sent her back home to Uganda to sample maize streak virus (see Chapter 7) and she returned with what is probably the most extensive such coverage ever undertaken.

Betty is currently a post-doctoral fellow working under the supervision of Prof Sir David Baulcombe, the Royal Society Professor of Botany at Cambridge University. She is working on sweet potato viruses, together with her alma mater, Makerere University. She first connected with David after she met a postdoctoral researcher from his Sainsbury Laboratory in Norwich, where he was the head before joining Cambridge. She visited him on her way home and managed to persuade him to allow her to come to his lab for a few months to complete some experiments we were having difficulties with in our lab. When she told me about this coup I was extremely impressed ... 'Where angels fear to tread'!

Florence Wambugu

During my frequent visits to Nairobi I met a truly remarkable woman, Florence Wambugu, who has single-handedly done more for agricultural biotechnology in East Africa than any other person I know. She was born one of nine siblings on a small farm in the Kenyan highlands and has childhood memories of going to bed hungry. She was fortunate to have a wise mother who sold the family cow to send Florence to school in the days when educating women was often considered a waste of money. She eventually became a plant pathologist specialising in viral diseases of potatoes and sweet potatoes. However, when I got to know her she was spearheading the use of tissue culture to propagate disease-free bananas.

It was typical of Florence to choose a crop that is considered 'a woman's crop'. In traditional African farming systems, women manage crops that feed their families while men are responsible for crops that make money. However, the man owns the land and ultimately decides on its use. So a woman's crop often suffers double blows, especially because, unless she can obtain credit, a woman must rely on her man to buy her planting material.

The traditional way of propagating new banana plants is to uproot a young sucker from close to the base of a mature plant. This is cheap but carries over whatever pests and diseases are present in the parent. Tissue culture involves the production of new plant material under sterile conditions in a laboratory. This also breaks the cycle of infection and can supply a 'hormonal kick' in the culture media, which results in higher yielding plants (Wambugu, 2001). But how was Florence to get these plants to women farmers?

She was on the board of the International Service for the Acquisition of Agribiotech Applications (ISAAA; see Chapter 2), and had close ties with the Kenya Agricultural Research Institute. Together they made funds available to launch the scheme, and the latest available information is that banana productivity has increased from 20 to 45 tons per hectare. This translates into an increase in income from a basic US$ 1 per day per family to as much as US$ 3 (www.absfafrica.org). Florence is currently the CEO of Africa Harvest, an organisation she founded.

Venturing into Nigeria

Around this time I was invited to a biotechnology conference in Enugu, Nigeria. I was just about the only white person, and certainly the only white woman, there. The opening ceremony was a real eye-opener to the type of life Nigerians lived under their dictator, General Abacha. The minister responsible for science and technology was due to open the meeting, but after an hour of waiting the local dignitaries went ahead with the ceremony in his absence. During the coffee break there came the sound of screaming tyres and in drove the convoy carrying the minister. Immediately everyone returned to their seats and the opening ceremony was repeated. However, the atmosphere was completely different and was set by the minister himself, also a general, who would not deign to try and adjust the microphone to his liking, but gestured to a minion who jumped to the task. The local dignitaries also adapted their speeches to suit his presence. After his departure I was approached by a TV interviewer who asked me what I thought about life in Nigeria. It was one of those moments when you watch your life pass before your eyes—just a few weeks previously my brother had been arrested for taking part in an illegal protest in Johannesburg, demonstrating against the death in custody of the Nigerian opposition leader, Ken Saro-Wiwa. I have to confess that I chickened out by saying I was there as a scientist and could only comment on biotechnology.

Fears around DNA transfer

In the late 1990s I was approached by the World Health Organization (WHO) to participate in a workshop on how GM crops could influence food security on the African continent. They asked me to address the question of the transfer of DNA from GM crops to bacteria and to mammalian cells, a talk which I later wrote up for publication (Thomson, 2001). There was a major argument at the time as to whether it was safe to use antibiotic resistance genetic markers as a tool in the selection of GM plants. Could the genes encoding such resistance be transferred from the food we eat into bacteria in our intestines, or indeed into the cells of our body, and thus result in ourselves becoming resistant to these antibiotics?

Gene transfer from a GM plant to a bacterium, or to humans or animals, is called horizontal gene transfer. It is the movement of genetic information between sexually unrelated organisms (different species). This is in contrast to vertical gene transfer, which occurs from parent to offspring. Let us consider the processes that would have to occur for horizontal gene transfer to take place from a food derived from a GM crop, the evidence for such processes occurring, and the possible consequences should they occur. I will use the example of a gene coding for resistance to an antibiotic.

Firstly the antibiotic resistance gene would have to remain intact in the gastrointestinal tract once the food is eaten. Enzymes in this tract degrade DNA into small fragments of about 500 base pairs, too small to encode antibiotic resistance within an hour. However, should a large enough fragment survive it would need to be taken up by bacteria or human cells. Although some bacteria can take up DNA from the environment by natural transformation processes, such transfers have yet to be shown by anaerobic gut bacteria in their natural environment.

What about uptake by human cells? All foods contain DNA and, although we have not accurately determined the amount that consumers ingest on a daily basis, estimates for cows indicate that they consume approximately 600 milligrams of DNA per day (Beever and Kemp, 2000). Any concerns regarding the presence of novel DNA in GM-derived foods must take into consideration that the DNA from this source would represent less than 1/250 000 of the total amount of DNA consumed. In addition, we are talking about one gene among the 20 000 to 40 000 genes found in a crop plant. In view of this and the ready digestibility of dietary DNA, there is an extremely low probability of transfer of antibiotic resistance genes from GM plants to mammalian cells. Indeed, the uptake of any exogenous DNA by mammalian cells is very difficult to prove.

I wrote this up in an appendix to *Genes for Africa* (Thomson, 2002: 187) and concluded by saying that, although the use of antibiotic resistance genes in transgenic crops is considered safe by scientists:

there is public perception that they could add to the already high levels of antibiotic resistance in pathogenic bacteria. Despite the fact that there is no scientific evidence to support this, scientists and regulators working in this field agree that they should use alternative transformation technologies that do not introduce antibiotic resistance genes in GM crops and foods.

Around the time that I was working on this I was also involved in meetings that were to have a profound effect on my future involvement with agricultural biotechnology in Africa. I had been invited to speak at the World Economic Forum in Davos.

Chapter 4

To Davos and further into Africa

The World Economic Forum, held in Davos, Switzerland every year, is an independent international organisation committed to improving the state of the world by engaging business, political, academic and other leaders of society to shape global, regional and industry agendas. It organises regional forums, and in 1999 I was invited to speak at the Southern African Economic Forum at the convention centre in Durban. I had no idea what to expect, but gave a run-through of my talk to a respected businessman I knew and he gave me lots of advice, which must have worked, because the next year I was invited to speak in Davos itself.

I was part of a discussion forum in which four speakers were asked to express their opinions on genetically modified crops. The audience was asked to vote before and after the presentations and there was a pleasing increase in the FOR vote at the end. But from my personal perspective, the most important interchange was between a member of the audience and Prof (now Sir) Gordon Conway, then president of the Rockefeller Foundation. 'Now that the foundation has successfully funded the development of genetically engineered vitamin A-enriched rice, what is their next target?' came the question. And his answer was, 'Drought-tolerant crops for Africa.' After the discussion I tapped Prof Conway on the shoulder and said, 'Gordon, can we discuss my research?' And so it happened that I received handsome funding from the Rockefeller Foundation from 2001 to 2006. But more of that later in Chapter 8.

I was invited back to Davos the following year and on that occasion my most significant experience came on the so-called 'free Sunday'. I had been asked to attend a meeting of the Informal Group of World Economic Leaders and on the bus discovered that my two co-speakers were Ian Wilmut of the Roslin Institute near Edinburgh, who had just cloned the sheep, Dolly, and George Church, a member of the

Human Genome Project. Our chair was Lord Robert May, president of the Royal Society of London. When I asked him what was expected of us he merely replied, 'Just respond to the questions.'

In my session I spotted the South African ministers for Trade and Industry (Alec Erwin) and Finance (Trevor Manuel). They kindly sat next to me and told me *sotto voce* who was asking the questions. Although we, the speakers, were introduced to the about 20-strong audience, we had no idea which world leader we were addressing. It turned out that they weren't that interested in Dolly, didn't quite realise the implications of sequencing the human genome, but certainly wanted to know more about GM crops. So I was in the hot seat for most of the two-hour session.

Davos taught me quite a lot about debating with the anti-GMO lobby. At the end of my first visit I found myself in a little makeshift TV tent together with Vandana Shiva, the Indian anti-GMO activist who opposes almost anything to do with big business (see Chapter 2). After one comment she made on poor Indian farmers being forced to buy seed against their will, I retorted, 'But that's a lie.' The interviewer merely smiled broadly, thanked us both and gave Shiva the last word on the subject. I realised that truth doesn't always make for a good interview. When I got home I drew up a fact and fiction table which I have found useful in similar situations (Table 4.1).

GM crops have continued to be on the agenda of World Economic Forums, especially recently with food shortage scares. In 2010 Bill Gates came out in support of the responsible use of the technology, especially for disease resistance and drought tolerance in developing countries. In 2011, governments of developing countries, together with NGOs and multinational food and agricultural companies, formed the task force 'Realising a new vision for agriculture'. However, the participants acknowledged that it would take not only innovation but a lot of collaboration to get things moving for technology-friendly regulations and infrastructure. At the 2012 meeting, Robert Carlson, leader of the World Farmers Organisation, noted that the world's agriculture industry had suddenly become one of the largest concerns among economic leaders of the world. He went on to say:

> Generally, at this forum, I feel the attitude on genetically modified action is if your culture will accept (it), and if it works for you and is safe, we encourage you to (accept it). The consensus among many of those attending is if we don't have genetically modified crops we won't be able to feed the world and will end up being even more dependent on using higher rates of fertiliser and pesticides (http://www.farmandranchguide. com/news/regional/carlson-representing-agriculture-at-the-the-world-economic-forum/article_56324758-50e0-b7cd-001871e3cebc.html).

Table 4.1: Facts and fiction about GM crops

	FICTION	**FACT**
	ENVIRONMENTAL IMPACTS	ENVIRONMENTAL IMPACTS
1	GM crops create superweeds	• The use of herbicides for decades has not resulted in superweeds • Herbicide rotation has been used for decades to prevent build-up of resistance
2	GM crops will destroy biodiversity	• GM crops are much easier to breed into different crop varieties as they only have one or a few linked genes added. Thus GM crops can increase crop biodiversity • Fewer insecticides are used leading to increased insect biodiversity
3	GM crops are harmful to the environment	• Insect-resistant crops have led to a decrease in the amount of pesticides used, leading to an increase in non-targeted insects • Some pesticides are also toxic to the humans who spray them. Insect-resistant crops prevent this harm • Herbicide-resistant crops have led to 'no till' or 'minimal till' agriculture. Instead of tilling the soil before planting to allow weeds to grow, spraying them with herbicides and allowing these to degrade before planting—leading to loss of top-soil—farmers now plant, allow the weeds to grow, and then spray. Result: soil improvement • Roundup, one of the herbicides that GM crops are resistant to, is rapidly biodegraded. Many conventional herbicides, such as atrazine, remain in the soil for longer periods
	FOOD SAFETY	FOOD SAFETY
1	GM foods are unsafe to eat	• No food in the history of humankind has ever been subjected to such rigorous safety tests as foods derived from GM crops • 2004: Food and Agricultural Organization 'no deleterious effects from consumption of foods derived from GM crops discovered anywhere in the world' • 2010: EU Commission Directorate for Research 'no new risks to human health or the environment from any GMO crops commercialized so far'

	FICTION	FACT
	MARKET ISSUES	MARKET ISSUES
1	GM crops are just a ploy of the multinationals to make more money	Farmers are savvy people. They will not buy seeds if they don't give them a profit. No-one is forcing farmers to buy seed from any given company
2	Farmers who plant GM crops have to buy seed every year	• Since the advent of hybrid crops/seeds in the mid-1920s, farmers who have chosen to plant such hybrids have had to buy seed every year. That was long before GM crops were even dreamt of • Farmers can choose not to buy hybrid seed but plant open-pollinated varieties, or land races. These have lower yields, but farmers can plant their own seed. These seeds are readily available from seed companies
3	GM crops cannot help to feed the poor	They could if they were allowed to be introduced. The developed world has imposed such strict regulations, which have to be followed by the developing world, that existing GM crops as well as new ones in the pipeline with improved nutritional content, and resistance to drought and disease, are extremely difficult and expensive to introduce
4	GM crops won't put more money into the pockets of smallholder farmers	• Currently, in 28 countries where GM crops are allowed, approximately 90 per cent are planted by smallholder farmers. Ask them why they buy GM seeds • In 2009, 87 per cent of the national Indian cotton crop was planted by smallholder farmers using GM seeds • In China the equivalent figure was 68 per cent • Smallholder farmers are the quintessential organic farmers as they cannot afford herbicides and insecticides; GM crops mean that they can improve their yield with seed alone, although addition of fertilizers will help
5	Genes can flow from GM crops and 'pollute' other crops	• Gene flow takes place between all crops, GM or non-GM. Conventional hybrid crops can just as readily 'pollute' local varieties

The African Agricultural Technology Foundation

In 2001 or 2002 I received an email from a fledgling organisation called the African Agricultural Technology Foundation (AATF), to be based in

Nairobi, asking if I would be prepared to have my name put forward as a member of their board. I submitted my CV and then forgot all about it. Some months later I was informed I had been appointed to the board, but I had completely forgotten what the initials AATF stood for. I checked on the Internet but all I could find was the 'American Association of Teachers of French'. Ashamedly I admitted my ignorance, and they kindly reminded me that this was an organisation funded by the Rockefeller Foundation, the UK Department for International Development (DFID) and the US Agency for International Development (USAID), whose aim was to transfer intellectual property in agricultural biotechnology from multinationals to African farmers.

I was unable to attend the first meeting, but discovered on my return home that I had been elected vice-chair of the AATF board. That was slightly unnerving, but nothing at all compared to the news received shortly thereafter that the chair had resigned and that I was now chair of the board of an organisation about which I knew extremely little. The interim executive director, Eugene Terry, kindly visited me in Cape Town to fill me in, but I was still pretty nervous when I came to chair my first meeting in Nairobi in 2003.

In 2004 the AATF was inaugurated on 16 June (my birthday), and was attended by representatives of the donors, many government authorities, partners and well-wishers. We had been given property on the campus of the International Livestock Research Institute (ILRI) in the rolling hills outside the city.

Figure 4.1   Cutting the ribbon with the Kenyan Minister of Agriculture, the Hon Kipruto arap Kirwa, at the launch of the AATF on 16 June 2004

One name stands out in the history of the AATF—that of Sir Gordon Conway. As the then president of the Rockefeller Foundation, one of the founding funders of the organisation, he was intimately involved in its conceptualisation and implementation. When he stepped down from that foundation he became involved with the UK Department for International Development (DFID), another founding funder. And then in 2009 he became a member of the board of trustees where he is chair of the development committee and an indefatigable fundraiser. His first book, *The Doubly Green Revolution: Food for All in the 21st Century* (Conway, 1997), was a source of inspiration for my second book, *Seeds for the Future*, for which he wrote the foreword. He has recently published *One Billion Hungry: Can We Feed the World?* (Conway, 2012) which should be required reading for all interested in food security. His continual emphasis on the importance of the AATF's end-users—smallholder farmers—and the improvement of their livelihoods keeps the organisation (and its board) focused on its primary aim.

Another important player is the current chair of the board, Professor Idah Sithole-Niang, from the University of Zimbabwe. Her term of office ends in 2013, but she has contributed a deep understanding of how academic knowledge can be translated into improvements on farmers' fields. She was also chair of the MSc in Biotechnology at her university when I was invited to be external examiner in 1992 (Chapter 3). This excellent course was dropped after a number of years, but Idah fought indefatigably for its reinstatement, and she finally won in 2012.

My involvement with the AATF started one of the most satisfying experiences of my life—helping to steer it to develop crops that could help smallholder farmers not only to become secure, but also to develop into commercial enterprises while, at the same time, helping African countries to become food secure. However, to fully understand the importance of the AATF I need to expand on the questions of intellectual property and the patenting of life forms.

Intellectual property and life forms

For centuries millions of intellectual property rights have been granted throughout the world under various IP laws in different countries for similar reasons: to encourage an inventor to disclose his or her invention to the public, thereby promoting the progress of science. This may be looked upon as a contract between a government and the inventor whereby the latter discloses the invention and the former provides the inventor with a monopoly for a given period of time, currently usually 20 years. This provides incentives for innovators to develop new technologies for that society.

Intellectual Property Rights (IPR) have revolutionised societies technologically, industrially and thus socio-economically. The ability of inventors to disclose their ideas in return for monopolies has facilitated the promotion of scientific progress. I recall hearing Ingo Potrykus, one of the developers of vitamin-enriched rice, so-called 'Golden Rice', speaking on the subject of patents at an international biotechnology meeting in Florida in 2002. He said he was initially outraged when he discovered that Golden Rice had made use of 70 IPRs belonging to 32 different companies and universities. However, he later realised that if these patents had not been in place, he would have had to have made those inventions himself in order to come up with his final product.

I met Ingo again at a meeting in Ravello, Italy in June 2012 and he was an extremely disappointed man. Opposition to Golden Rice, led largely by Greenpeace, had prevented its uptake. Golden Rice contains beta-carotene, which is converted into vitamin A in humans. According to the WHO, between 250 000 and 500 000 children become blind every year due to vitamin A deficiency. Half of these children die within a year of going blind. Ingo Potrykus and colleagues developed Golden Rice in 1998, but its introduction has been blocked by Greenpeace who claim that there are better ways to alleviate this deficiency, such as vitamin A pills and 'home gardening'. Yet they are doing nothing to implement alternative programmes for the millions of victims and, in the words of Dr Patrick Moore, the co-founder of Greenpeace, this constitutes a 'crime against humanity' (Moore, 2012).

Products and processes involved in GM technology are patentable under the international Union for the Protection of New Varieties of Plants (UPOV). Such patents have been viewed with great hostility in general from detractors of GM crops, including most southern African countries, with the exception of Kenya and South Africa, which have been members of UPOV since 1978. The arguments against this system of protection include the view that it is excessively monopolistic and protects the breeder to the disadvantage of farmers' rights in indigenous knowledge. Kenya, however, saw the advantage of new plant varieties in horticulture. Access to quality seed and horticultural material, such as flowers and vegetables, has facilitated global trade in these commodities (Olembo, 2008) and Kenya is now a major exporter of cut flowers and fresh vegetables. However, the most significant impact of a plant protection system is its stimulation of research in agricultural productivity.

Many of the arguments against IPR in African agriculture are spelled out in a paper on food security by Kuyek (2002). His arguments include the threat of monopolies by multinational companies, placing needed products beyond the reach of poor countries. There are also fears that patents threaten the freedom of farmers to access seed. Indeed it is true that global corporations hold 90 per cent of all technology and product patents related to living materials. But that is partly due to the expense involved in developing and bringing such products to market. It is estimated that to develop and bring a GM crop to market costs US$136 million, of which 26 per cent of total cost and 37 per cent of time (an average of 13 years) are involved with regulatory expenses (Phillips McDougall, 2011).

I noticed the swing away from the involvement of publicly funded institutions, such as universities and research institutes, towards private corporations, as early as 1990 during meetings of the International Symposium on Genetics of Industrial Microorganisms, long before the advent of GM crops. Whereas in the early days of the symposium, colleagues at universities and research institutes readily shared their research results in an open forum, as time went by this happened less and less often. The podiums became dominated by scientists from private companies who would share data only on products and processes that had already been patented. Research is expensive and requires considerable time—an estimated 13 years from the start of a GM crop to its appearance on the market. It requires the use of skills and costly equipment that push up the value of the final product. Compensation for such involvement becomes a necessity, and securing IPRs provide such a mechanism.

During the October 2002 World Summit on Sustainable Development in Johannesburg, heated debates occurred in various forms on the ills of IP as a medium for trade (Olembo, 2008). Claims were made that the multinational seed industry's expansion into Africa had come with intense pressure in favour of patented products, but with no intention of making the technology freely available to farmers. Some views expressed at this meeting were that African agriculture does not require IPR because such agriculture is led by farmers, funded by the public sector, and based on collective knowledge. Anti-IPR activists claimed that protection regimes undermine farmers' rights, foster dependence on foreign companies, allow piracy of farmer-developed crops, and threaten food security and biodiversity.

The contrasting views were that because of the need to increase productivity, the situation in Africa is no longer static; it is evolving all the time. Local companies, national research institutions, non-governmental organisations and farmers' associations are increasingly engaging in biotechnology and other improved agricultural techniques for higher agricultural yields (Olembo, 2008).

Both of these views have validity and somehow must be harmonised. In fact, as Klaus Amman points out in his article entitled 'Reconciling traditional knowledge with modern agriculture: a guide for building bridges' (Amman, 2007), the barrier between these two approaches is artificial. Many scientists depict traditional knowledge as closed to conceptual inputs from outside, whereas science is open to new thought, precise in its empirically tested progress. Critics of science, however, mistrust it for being too abstract, analytical and divorced from the needs of real people. The reality in both cases is different from these perceptions. Traditional knowledge that has accumulated since ancient times and been transmitted by oral tradition has often turned out to be strikingly precise when tested against empirical observation. Indeed, given the test of time, traditional knowledge is verified or falsified by experiment and observation. And in Western science, oral tradition is certainly present. Scientific communities with different views and lexicons continue to exist regionally despite the homogenising influences of the scientific literature and the Internet—for instance in botanical and ornithological nomenclature.

However, the fact remains that in order for Africa to benefit from these new technologies, and not to be left behind as it was in the Green Revolution of Asia in the 1960s and 1970s, African countries will have to develop policies to deal with IPR. Apart from straightforward negotiations between potential African users and IPR owners, in which IP may be acquired through licensing, outright purchase or partnerships and the need to minimise costs, particularly to deserving poverty-stricken developing countries, may require goodwill arrangements. It was with this in mind that the AATF was established.

How the AATF works

The rationale for the establishment of the AATF is spelled out in a paper entitled 'The African Agricultural Technology Foundation approach to IP management' (Boadi and Bokanga, 2007). The fundamental reasoning was to establish links between private and public sector institutions owning technological innovations in developed countries, and African stakeholders in agricultural development, such as the National Agricultural Research Services, farmers' associations, non-governmental organisations, and national, private sector agribusinesses, such as seed companies. The goal of the AATF is to facilitate access to appropriate scientific and technological resources, whether proprietary or not, and to promote their adaptation for use in specific projects intended to increase the productivity of smallholder, resource-poor farmers in sub-Saharan Africa.

In order to do this effectively, the AATF first consults with these African stakeholders in order to identify what the priority crops and the key constraints are for resource-poor farmers. They then consult with potential technology providers, in private and public sectors, to identify relevant technologies that can address these constraints. They must then negotiate with potential partners to develop a project business plan that specifies the role of each partner institution and determines how and where the technology will be used. The AATF then enters into licensing agreements to access and hold proprietary technologies and ensure freedom to operate for all components of the technologies.

Once all this is in place, work proceeds by sublicensing partner institutions to carry out the actual research as needed to adapt the technologies to smallholder farming conditions in Africa. In some cases this will require the gene(s) of interest to be transferred to an African crop, such as cowpeas or African varieties of crops such as maize. These must be tested for regulatory compliance before the products can be produced and distributed to farmers. Once this is done, the AATF must put in place systems to monitor compliance with the regulations and to minimise the risk of technology failure. It further facilitates the work of appropriate partner institutions to ensure that links in the value chain are connected, are effective, result in technology products that reach farmers and, most importantly, allow farmers' surplus harvests to reach markets. I recall asking some farmers during an AATF field visit what their major problems were and they identifed 'drought' and 'markets'. To ensure that these efforts are sustainable in the long term, the AATF also creates partnerships within African countries and with external stakeholders to develop the necessary indigenous capacities.

It is clear from this that the AATF operates along the entire product value chain, from the transfer and adaptation of technology to farmers' access to output markets, with each implementation step undertaken with the relevant partner organisation. The nature of the AATF's involvement varies from project to project, depending on the specific requirements and issues that need addressing (see Figure 4.2).

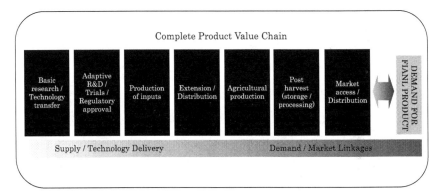

Figure 4.2 AATF's involvement in the complete product value chain

IR maize

I will illustrate the AATF'S process of embarking on projects using a few examples (for further details see www.aatf-africa.org). Our first product was maize resistant to the parasitic weed, *Striga*, commonly called witchweed. It is an extremely pretty, but insidiously dangerous plant (figures 4.3 and 4.4) which relies on maize for about 40 per cent of its nutrients. It sets millions of tiny seeds which can lie dormant in the soil for years until maize plants start to germinate. Chemicals released into the soil then send signals to the *Striga* seeds that food is about to become available. The seeds in the vicinity of the developing maize plants germinate and infiltrate their roots into the roots of their hosts, eventually smothering the plant and effectively preventing farmers from getting rid of these intertwined weeds by physical means. Fields infected with *Striga* seeds are left unplanted by maize because farmers realise it is useless even to try to plant the maize there. The company BASF developed a variety of maize tolerant to the herbicide Imazapyr— the so-called Imazapyr resistant (IR) maize. They did this not by genetic engineering but by classical breeding, and therefore AATF's first product was not a GM crop. This meant that its distribution was unhampered by the stringent requirements for the deployment of such crops and enabled us to develop it much more rapidly than we would have been able to, had it been a GM crop.

Figure 4.3   A field of maize damaged by *Striga*

Figure 4.4   The difference between cobs derived from Imazapyr-resistant maize (left)
and sensitive (right) maize

During one of our board meetings, members flew to western Kenya to view some of the field trials. The only reason we had been successful in distributing the IR seed to farmers was due to our involvement with local organisations such as WeRATE, a consortium of NGOs, community-based and farmers' associations. It was thanks to them that we were able to stem a potential problem. We had distributed the IR maize seeds coated with the herbicide, Imazapyr. But many farmers sow seeds by hand and intercrop maize with plants such as legumes, so if they planted the legume seeds without first washing their hands the legumes would have become covered in Imazapyr. Imagine the headlines in *The Nation*: 'Biotech maize kills beans!' The AATF rapidly brought out simple brochures in all the local dialects and, with the help of the NGOs, distributed them widely, thus preventing what could have been a minor disaster. Now Kenya Seed Company provides two pairs of gloves with each package of seed, together with illustrations on how to handle the seed.

However, the biggest problem with this project has been the production of seed. Even today, some seven years after the launch of the *Striga* Control Project, Dr Gospel Omanya, the AATF seed systems manager, bemoans the slow output of seeds. This is to some extent understandable. The production of these seeds requires that they be coated with Imazapyr which, in turn, involves investment in costly coating equipment. Production of the seeds has increased, even so. During 2010, 30 tonnes of seed were produced in Kenya, up from the 20 tonnes in 2008, and could be distributed to farmers through the agro-dealer network in western Kenya. The project is being taken up across Africa. Tanzania has begun production but the herbicide still has to be registered there. In Uganda, on-farm variety trials have been carried out to collect data for performance trials, because varieties that work in Kenya may not be suitable for the climatic, soil and other conditions prevailing in Uganda. Nigeria is working towards commercial release. Regional trials in Zimbabwe have yet to identify their best adapted lines.

Bt cowpeas

A second AATF project example, which does involve genetic engineering, is the *Bt* cowpea project. *Bacillus thuringiensis* is a naturally occurring soil bacterium which produces proteins, called *Bt* proteins, that are toxic to certain insects. They cause little or no harm to most non-target organisms, including humans and wildlife. They have been used in sprays in conventional and organic agriculture for decades with negligible or no ill effects on the environment or human health. Thus, *Bt* toxins are considered an environmentally friendly alternative to broad-spectrum insecticides. From the mid-1990s, crops expressing *Bt* genes and hence producing the toxins inside the plant, have been commercialised in the US and *Bt* crops such as cotton and maize have been planted commercially in South Africa.

Cowpea (*Vigna unguiculata L. Walp*) is considered the most important food grain legume in the dry savannahs of tropical Africa, where it is grown on more than 12.5 million hectares of land. It is rich in high-quality protein and has an energy content almost equivalent to that of cereal grains. It is a good source of quality fodder for livestock and provides cash income. Nearly 200 million people in Africa consume it. However, many biotic and abiotic factors greatly reduce cowpea productivity in the traditional African farming systems. Among these constraints is the pod borer, *Maruca vitrata*, which perennially damages cowpea pods on farmers' fields. In severe infestations, yield losses of between 70 and 80 per cent have been reported. Control through spraying with insecticide has not been widely adopted by farmers due to the prohibitive costs and health hazards associated with spraying (www.aatf-africa.org).

The *cry1Ab* gene, coding for one of the many *Bt* proteins, was obtained from Monsanto, and a group at the Commonwealth Scientific and Industrial Research Organisation (CSIRO) in Australia, led by Dr TJ Higgins, transferred it into cowpea. Promising transgenic events were selected after thorough laboratory and greenhouse testing. These were subjected to confined field trials (CFT) in 2008 in Puerto Rico, chosen because that country had the required biosafety regulations in place. One of the lessons learned there was that natural infestations by *Maruca* could not be relied upon. Indeed, the insect population that season was almost too low to evaluate the effectiveness of the transgenic lines. CFT began in Nigeria in 2009 in partnership with their regulatory authority and the ministry of environment. Again, one of the greatest challenges has been in obtaining significant *Maruca* infestations in the field. To solve this, AATF partnered with the International Institute of Tropical Agriculture (IITA) in Ibadan to train entomologists and technicians to rear the insects in the laboratory and then artificially infest the larvae into the CFT site. 'We were delighted to note that while the larvae fed on the non-transgenic cowpea, they did not attack the transgenic varieties,' said Dr Misari, the project entomologist (AATF Annual Report, 2010). The project has had excellent cooperation from the Nigerian authorities to the extent that the AATF has established an office there.

An interesting connection transpired in relation to this project. Ed Southern, who had taught me to use Southern blots at the 1978 course I attended in Basel (see Chapter 1), had made a lot of money, not out of his blotting technique, which I daresay he never thought to patent, but out of another major technology, that of DNA microarrays. You'd think it would be enough to invent one technique that would revolutionise molecular biology, but two ... ! He had started the Kirkhouse Trust, which focuses, inter alia, on agricultural crop improvement research for the developing world, specifically legumes. He decided to support the AATF's project on insect-resistant cowpeas in West Africa, and on hearing of my involvement, paid a visit to the University of Cape Town while visiting Ghana (not everyone is aware of distances in Africa) to inform our students of the importance of this work. So we met up again after some 30 years, under very different circumstances, but still with the same aims in mind.

Banana bacterial wilt

A third AATF project is targeted at diseases of bananas. Bananas and plantains are an important food source for more than 100 million people in sub-Saharan Africa. In the East African highlands and most of the Great Lakes region, bananas are a major staple food and a source of income for over 50 million smallholder farmers. East Africa produces 16.4 million metric tonnes of this crop per year, about 20 per cent of the world output. However, many biotic and abiotic factors greatly reduce productivity for bananas cultivated under traditional African farming systems. For instance, in 2001, an outbreak of banana bacterial wilt, caused by *Xanthomonas campestris* pv *musacearum*, broke out in Uganda, leaving in its wake a trail of crop destruction and utter misery among affected farms. It later spread to the Democratic Republic of Congo, Rwanda, Tanzania and Kenya, and is very destructive, infecting all banana varieties. The International Institute of Tropical Agriculture (IITA) estimates economic loss due to diseases in Uganda alone to be at a staggering US$ 200 million. The AATF is collaborating in a public/private sector partnership project to develop banana bacterial wilt-resistant transgenic bananas in East African-preferred germplasm (www.aatf-africa.org).

The genes involved are either those coding for the sweet pepper ferrodoxin-like protein (Pflp) or the hypersensitivity response assisting protein (Hrap). The AATF is collaborating with scientists from the Academia Sinica in Taiwan, who showed that these genes improved disease resistance of vegetables, including broccoli, tomatoes and potatoes. The AATF received a royalty-free licence to use these genes in 2006. On 5 October 2010, the AATF, the IITA and the National Agricultural Research Organisation (NARO) of Uganda planted transgenic lines in a confined field trial in Kawanda, in south-east Uganda (AATF Annual Report, 2010). It attracted an item in the journal *Nature*, in which Linda Nordling wrote, 'The new variety is part of a wider effort to improve the East African Highland banana, a fruit so important to Ugandans that its name, matooke, is synonymous with food in one of the local languages' (Nordling, 2010). The project partners plan to grow the resistant bananas in five countries in the Great Lakes region: Kenya, Uganda, Tanzania, Rwanda and Burundi.

Bearing in mind the lessons learned from the *Striga* project, where seed propagation has been a problem, the partners in this project are working to ensure that the practitioners on the ground are well prepared and skilled for the mass micro-propagation in tissue culture, and the dissemination of banana plantlets in the target countries.

This exercise marked a major step for the project partners towards addressing some of the legal roadblocks regarding deployment of GM crops in Uganda. In 2010, Uganda's biosafety law only existed in draft, based on a biotechnology and biosafety policy adopted in 2008. However, AATF, IITA and NARO worked in partnership to ensure compliance to the regulatory requirements and approval by the Uganda national biosafety committee. It is hoped that this will in time lead to the government passing a Biosafety Act. If not, the words of Linda Nordling in her 2010 article in *Nature* may come back to haunt the project: 'Delays to a law regulating the commercial growing of genetically modified food in the country means it is not clear when the improved banana could be released to farmers.' (Nordling, 2010)

Another extremely exciting AATF project involves GM maize developed to protect plants against drought. However, this water-efficient maize for Africa (WEMA) will be dealt with in Chapter 8.

A major concern of the AATF's project collaborators, whether they are public entities or multinational companies, is liability exposure once proprietary technologies have been licensed to the AATF and subsequently sublicensed to other parties for use in sub-Saharan Africa. A related concern is the possible misuse of the technology and associated confidential information. The AATF has developed a proactive product stewardship mechanism to address these concerns. It ensures that smallholder farmers and research partners comply with all relevant licensing conditions and regulatory requirements.

It further protects technology donors from liability through indemnification provisions and warranty disclaimers in agreements and by conducting a comprehensive risk analysis for each project (Boadi and Bokanga, 2007).

The farmers' view

One day, I received an email from David Hoisington at the International Crops Research Institute for Semi-Arid Tropics (ICRISAT), asking me if I would serve on a panel to undertake a review of the institute, which is based at Patancheru, near Hyderabad in India. We met in Nairobi, where ICRISAT has a regional hub, and visited its research centre on the road to Mombasa. There we saw its five main crops—chickpea, pigeon pea, sorghum, ground nuts and pearl millet—in breeding trials. We also visited a number of farms. On one of them a woman farmer, dressed in ragged clothes but carrying a cell phone, showed us the difference in yield of the local varieties (landraces) and ICRISAT's improved lines. She told us that she would never plant landraces again. David and I couldn't help smiling at each other—so often in public meetings those of us advocating the planting of GM crops are accused by opposition activists of forcing local smallholder farmers to stop planting landraces.

Another interesting farming ploy she explained to us was the way in which she deals with the problem of birds eating her sorghum and pigeon peas, especially during their annual migrations. She intercrops them with maize, which grows taller and forms a protective canopy during the bird migrations. Afterwards, having harvested the maize, her sorghum and pigeon peas are free to mature.

After Nairobi our team split up to visit the various stations in Africa, and David and I went to Lilongwe in Malawi. The staff took us to meet the local NGO, the National Smallholder Farmers' Association of Malawi (NASFAM), who dealt with both the growing and the marketing of farmers' crops. This was a real eye-opener. The NASFAM base we visited was equipped with a large warehouse to which farmers brought their groundnuts to be weighed and graded. Depending on the grades of the nuts they were either rejected or accepted for local, South African or international markets. The grades are based, to a large extent, on the presence of aflatoxins, which are harmful toxins caused by the infection of certain types of fungi. The testing for aflatoxins is done initially at ICRISAT's labs and later in labs in South Africa. One of the reasons for the infection is that, after harvesting, the encased nuts are soaked in water to ease the hulling process. Wet nuts, lying under the hot African sun, lead to a perfect breeding ground for the aflatoxin-producing *Aspergillus* fungi. We were taken to a village where a woman farmer explained to us her farming practices. I asked if she could tell us what she understood about aflatoxins, their causes and effects. Her answer would have made any of my undergraduate students proud.

Addressing the UN

Another useful opportunity to raise the issue of GM crops in Africa came in 2002, when I was asked if I would address the United Nations as the guest of the Secretary General, Kofi Annan. What a question, but what to wear? I was in Entebbe, Uganda and in the airport on my way home I spotted a rather splendid-looking outfit outside a souvenir shop. I tried it on over my shorts and T-shirt and reckoned that with a bit of alteration it would do. Well it did more than 'do'—when I met Kofi Annan he said, 'Wow, you are an African!'

Kofi Annan, or SG, as his adoring staff called him, had recently instigated a series of lectures for the United Nations ambassadors and staff. I was speaking together with a colleague from Chicago, and ours was the second in the series. We arrived to have a run-through of our presentations in the morning to discover that the auditorium was available for only a short time and we would not be able to check our talks via the projector, merely load them onto the computer.

We were ushered into lunch in the SG's private dining room on the top floor of the UN with stunning views over the East River and the Chrysler Building. The wine steward came round and I at first declined but Kofi Annan quietly told me he thought it would do me good. As the lunch drew to a close I began to look nervously at my watch. The SG noticed and, patting my hand, said, 'Don't worry, nothing will happen until I get there.' That wasn't quite the point. I was rather concerned as the IT set-up in the general assembly chamber was less than ideal and there had been no time to see how my presentation, prepared on a PC, would appear on an Apple Mac. Wait and see was all I could do. However, my fears were for naught and all went well.

It was an open seminar series in which the ambassadors would not be sitting behind their country names. One of the first to make a comment was the ambassador from Zambia. Earlier that year, when people in this country were on the verge of starvation due to a prolonged drought, the Zambian government had banned the importation and distribution of food aid as it might contain GM maize. President Mwanawasa, in defending this stance, said that this maize might be toxic and, in addition, if farmers planted it, instead of eating it, this could jeopardise the country's non-GM status and thus its maize exports. (This, in spite of President Thabo Mbeki's offer to have all the maize milled in South Africa, with only maize meal being sent to Zambia.) When an opposition member of parliament, Vitalis Mooya, challenged this view and reported that three elderly women had died of hunger and villagers had resorted to eating toxic roots in the Southern Province (*Business Day* 15 October 2002) he was arrested and interrogated for his trouble.

In front of the UN ambassadors, I tried to explain to the Zambians that there was no evidence that GM maize was toxic and even if farmers were to plant the seeds the maize would inevitably die after a few weeks due to African diseases such as maize streak virus to which maize from other countries is supremely sensitive. But I am afraid my words fell on deaf ears. Indeed, I was later told that when a delegation of Zambians visited America to discuss this issue, their scientists were not allowed to meet American scientists without the presence of politicians. In addition, among the stories the Zambians had been told by anti-GM activists in Scandinavia was that if men ate GM maize they would become sterile.

Years later, at a meeting in Rome in 2011, I heard the origin of this story from Marc van Montagu. Apparently, some scientist had fed mice GM maize and claimed to have evidence that some of them showed decreased fertility. However, despite the work never having been corroborated, it was picked up by the press and a photograph was taken showing how the news had been interpreted and spread (Figure 4.5).

In 2012 I attended a meeting organised by the African Development Bank in Nairobi, where I heard a politician from Zambia saying that his government had changed its attitude to GM crops and were now keen to allow field trials to test their efficacy in his country. I doubt it had anything to do with my address to the United Nations but it would be nice to think it might have played some little part!

L'Oreal/UNESCO award

In 2004 I received the L'Oreal/UNESCO award For Women in Science for Africa. This is an annual award which aims to improve the position of women in science by recognising outstanding women researchers who have contributed to scientific progress. Each year the award

Figure 4.5   Protesters in India: GM foods will make you sterile
(Source: Greenpeace, India)

alternates between life and material sciences, and an international
jury selects a winner from each of Africa and the Middle East, Asia-
Pacific, Europe, Latin America and the Caribbean, as well as North
America. Pascale Cossart, who had been on the same EMBO course I
had attended in Basel in 1978, was the first European winner in 1998

and was on the jury that chose me. Valerie Mizrahi, who had worked in my Laboratory for Molecular and Cell Biology in Johannesburg in the 1980s, was the African/Middle East winner in 2000. And in 2012 the winner for this region was Jill Farrant, with whom I share an office and a lab at UCT.

I was given the award 'for work on transgenic plants resistant to drought and to viral infections, in an effort to respond to the continent's chronic food shortage'. This was a tremendous honour and I was impressed that the judges would vote for someone working on GM crops, knowing the antagonism towards this field in France, where both L'Oreal and UNESCO are based. Indeed years later I learned that the jury had indeed considered the potential negative fall-out such an award might have.

In the run-up to the awards, L'Oreal sends a team of video and stills photographers to prepare promotional material on each of the winners. Mine arrived just before Christmas in 2003 when, unfortunately for South African farmers, the country was in the grip of a crippling drought and maize seedlings on the country's farms were taking a severe beating (Figure 4.6). However, this was fortunate for the team, as one of the projects for which I received the award was the development of maize tolerant to drought (see Chapter 8).

Figure 4.6 With Andries Botha on his maize farm in Viljoenskroon, Free State

The ICGEB project

The first PhD student to work on the drought-tolerance project was Dahlia Garwe. She had been a member of the MSc Biotechnology class at the University of Zimbabwe I examined in 1992 (see Chapter 3) and the third of four PhD students I supervised from that country. She, alone among them, had returned to Zimbabwe, where she had resumed her position at the Tobacco Research Board (she is currently their acting managing director). In 2007 the International Centre for Genetic Engineering and Biotechnology (ICGEB) put out a call for research projects on drought tolerance for developing countries. There are 60 ICGEB member states, of which South Africa is one, and this call required scientists from these countries to submit proposals together with one or more developing countries. I submitted one together with Dahlia and a colleague from Kenya and we were among the successful candidates asked to submit a more detailed proposal. It would require the work to be done at three institutions in Zimbabwe, Kenya and South Africa

Before embarking on this phase of the project I thought it would be prudent to visit Dahlia's institution in Zimbabwe and make sure they were able to deliver, because Zimbabwean finance was in rather a parlous position. When I arrived in Harare, I discovered that Dahlia's family hadn't had bread for three days and there was no water in the municipal supply to the area of Kutsaga, where the Tobacco Research Board is. Many of the shops were empty and I heard Dahlia on the phone to a family member arranging an exchange of cooking oil for other items. When I asked what she put into her children's lunch boxes, she said even if they did have bread there was only a bit of jam to put on sandwiches. But the hotel I stayed at was like any 4-star hotel in Europe or elsewhere, with the breakfast buffet overflowing with bacon and eggs (Dahlia hadn't seen any of these in months).

Even so, in the Tobacco Research Board laboratory work was on the go. Students and technicians were at their benches, the autoclaves were being used, despite the fact that water had to be carried in by hand from Kutsaga's local supply, the fridges and deep freezes were well stocked with reagents and the tissue culture facility was busy. It became clear, however, that although it was called the Tobacco Research Board, not much tobacco was coming in as so little was being produced, with so many farms no longer functioning.

I left Zimbabwe feeling reasonably optimistic that we could deliver if we received the ICGEB grant. In due course, our project was approved and we set to work. The first item on the agenda was recruiting a PhD student from each institution to begin work at UCT. My Kenyan colleague found a suitable one working at KARI, the Kenya Agricultural Research Institute, but Dahlia had problems as her chosen student decided to leave Zimbabwe for more promising pastures. She did manage to identify an MSc student and fortunately the ICGEB accepted this compromise. However, problems raised their heads soon after the two students arrived.

Their hands-on and academic backgrounds were not up to the required standards, putting a heavy burden on the other members of my lab, who had to spend long hours showing them the various techniques. The new students also appeared to have unrealistic expectations of what the ICGEB grant meant for their own personal finances and I soon found that I was dipping into my other open research funds to bail out the project. After about nine months I reluctantly informed the ICGEB that I wished to terminate the project. The relief shown by members of my lab and our administration office who were handling, among other things, the finances of the project, made me realise what a burden I had placed on them. The moral of this sad story is: always choose your own postgraduate students. My African colleagues were both very understanding as they, too, had not had first-hand experience of the students and had just hoped they would work out alright when the students were immersed in my lab. This is in stark contrast to my excellent experience of supervising PhD students under the USHEPiA programme. There, great emphasis is placed on meetings between students and supervisors in both participating countries well before any research takes place.

The InterAcademy Council

Another organisation I became involved with was the InterAcademy Council (IAC), an arm of the InterAcademy Panel that represents all the world's scientific academies, such as the Royal Society of London, the National Academy of Sciences in the US, the Indian National Science Academy, Science Council of Japan, and so on, and includes the Academy of Science of South Africa, of which I had been one of the founding vice-presidents. The IAC had been asked by Kofi Annan to undertake a study titled 'Food for Africa: harnessing science and technology to increase agricultural productivity in Africa'. We were chaired by Speciosa Kazibwe, Vice-President of Uganda and a former minister of agriculture, and MS Swaminathan, a former secretary of agriculture in India and a winner of the prestigious World Food Prize.

One of our recommendations was to bridge the agricultural genetic divide between African countries and those in the developed world. This would need substantial investment to respond to the specific needs of African farmers if they are to derive benefit from both conventional breeding and biotechnology. Technology needs to be fine-tuned to African needs, one of which is the dominance of weathered soils and their concomitant poor fertility. We noted the long gestation period biotechnology requires before its impact can be realised and urged for investment sooner rather than later.

In 2010 I was asked by the Canadian International Development Research Centre (IDRC) and the Canadian International Development Agency (CIDA) if I would serve on a committee of assessors to judge applications for projects on food security. This was a new programme aimed at linking Canadian research institutions with those in developing countries and was being jointly managed by the CIOA and the IDRC. The first call resulted in more than 300 applications, far more than they had expected. Of the finally approved 13 projects, with an average value of CA$ 2.5 million and expected to produce results by the time the projects end in 2015, some included the use of biotechnology in improving vegetables and pulse crops in Nigeria and Ethiopia.

One of the most exciting outcomes of all the experiences I have had resulting from my exposure to the international biotechnology community is how they all come back to the importance of this technology for Africa. However, all the technology in the world will not help hungry Africans feed themselves unless the authorities in each country are willing to accept that, as with all new technologies, there are risks that must be balanced with the potential benefits. Of course, and this is something which is often forgotten, it is equally important for those authorities to weigh the costs to their people of not accepting a new technology that could help to solve the problems of food insecurity, especially in times of climate change. The next two chapters will deal with these conundrums.

Chapter 5

A South African National Biotechnology Strategy

In the early days of South Africa's new democracy, a foresight board was established to look into the directions the country should be taking in technology research and development. During that exercise the chair, Rob Adam, said that the most startling innovations would occur at the confluences of three profound scientific currents: quantum mechanics, information technology and biotechnology. If we wanted to be a competitive country and, indeed, a competitive continent, we needed to ride these waves and to know as best we could where they would take us (A National Biotechnology Strategy for South Africa, 2001). The foresight study indicated that developments in information technology and biotechnology would be the cornerstone of the knowledge-based economy.

In due course, the then Department of Arts, Culture, Science and Technology (now the Department of Science and Technology) set up an expert panel of 10 which assembled in May 2001 and, over a period of two weeks, drafted a National Biotechnology Strategy. During that intense period, when most of us lived together, eating, sleeping and drinking biotechnology, we interviewed 30 people including representatives from industry, finance, government and even the anti-GMO lobby. Prof Iqbal Parker of the Health Faculty at UCT was the chair. Other members included Prof Frikkie Botha, an agricultural biotechnologist; Dr Winston Hide, an IT specialist; Dr Mohammed Jeenah, an agricultural scientist; Dr Nozibusiso Madolo from the National Department of Health; Prof Mbudzeni Sibara, a microbiologist; Dr David Walwyn, a technology manager from the CSIR (our scribe); Prof Brenda Wingfield, a forestry pathologist; Dr John Mugabe, a technology expert; and myself. As our group was almost entirely drawn from the academic sector, we took great care to interview as many people from the private sector as we could.

Our report, which was published in June of that year, made three major recommendations:

(1) Biotechnology regional innovation centres (BRICs) should be established.

(2) A National Biotechnology Advisory Committee should be established.

(3) The government should articulate a single policy position on biotechnology.

It seemed to take ages but, at last, the Biotechnology Strategy was approved by cabinet. A few weeks later I was at the celebrations in parliament for President Thabo Mbeki's 60th birthday and, when I was introduced to him by Thoko Didiza, the Minister of Agriculture, I rather impolitely said 'Mr President, it's all very well for the Cabinet to pass the Biotechnology Strategy, but where's the money?' With his usual aplomb he assured me that biotechnology was very important for South Africa's economy and it would be funded. I doubt whether my intervention played any part in subsequent events, but in due course four biotechnology regional innovation centres (BRICs) were indeed established.

The Biotechnology Strategy had recommended that the centres be regional. And indeed, three of them were, concentrating largely on biotechnology issues that were strong in their geographical regions, such as mining and veterinary science in Gauteng. However, when PlantBio was established it became clear that these activities were in no way concentrated on any particular region, but were countrywide. Hence the acronym, BRIC, became BIC. The National Bioinformatics Network also became a component of the BICs.

The biotechnology innovation centres

The first BIC was LIFELab, and it was decided to base it in Umbogintwini, KwaZulu-Natal, just south of Durban, because this was in close proximity to the SA Bioproducts facility. This facility was started in 1993 as a joint venture between African Explosives and Chemical Industries (AECI) and the Industrial Development Corporation (IDC) in order to produce the amino acid lysine. They chose Umbogintwini because they already had an established manufacturing infrastructure there. It had ready access to its source of raw material—sugarcane—and was close to Africa's largest harbour. AECI had diversified in the 1980s into amino acid production and they opened the lysine plant in the 1990s. During

1998/9 the company underwent restructuring and SABioproducts was established as a management buyout of the lysine production facility. Later, this company diversified into other amino acids and were thus the perfect partner for LIFELab. Apart from their liquid fermentation activities for the cultivation of animal, bacterial and plant cells, LIFELab concentrated on infectious diseases such as HIV/Aids, TB and malaria, as well as bioprocessing.

A second BIC was BioPAD, based in Gauteng, which began with a very broad portfolio of projects covering mining, environment, industry and animal health, but they soon narrowed it down to health biotechnology. Their focus was on diagnostics, drug discovery and development, vaccines and bioprocess technologies.

Cape Biotech Trust also concentrated on healthcare products, but in addition funded a start-up company, Cape Carotene, which produces astaxanthin, the pigment that makes salmon and flamingos pink. It uses algal technology developed by members of the UCT Department of Chemical Engineering.

The fourth BIC was PlantBio, based in Pietermaritzburg in KwaZulu-Natal, focused, as the name implies, on plant biotechnology. At its launch, Minister Mosibudi Mangena of the Department of Science and Technology was due to officiate. The PlantBio CEO, Sagadevan Mundree (the same Saga who played an important role in my lab's development of drought-tolerant maize; see Chapter 8) introduced me to him very early in the informal part of the proceedings. When I told the minister that I was a member of his National Advisory Council for Innovation his response was that I was obviously not doing my job as I hadn't advised him on biotechnology! I proceeded to correct that omission and during our discussion I mentioned that if South Africa was serious about developing an agricultural biotechnology business sector it would have to do something about the shortage of skills in the areas of plant physiology and plant breeding. To my surprise, he touched on this during his speech. Unfortunately, however, these shortcomings still exist in the scientific landscape of South Africa.

The study of plant physiology is becoming rarer and rarer at South African universities. My research group is extremely unusual in having (and very fortunate to have) Jill Farrant, one of South Africa's leading plant physiologists and, as mentioned in the previous chapter, the winner of the 2012 L'Oreal/UNESCO Women in Science for Africa, as a member of our team. But she is in the Department of Molecular and Cell Biology and not in the Department of Botany. Unlike plant molecular biology, plant physiology is not seen as 'sexy' by university students. Students of plant breeding are also becoming rare commodities. This is perhaps understandable as the nature of plants is to grow rather slowly, and to complete a three-year (or, more usually, four-plus year) PhD on this topic is not easy. Most plant breeders work in the private sector and it is therefore also difficult for potential PhD students to find academic supervisors.

During its short life, PlantBio helped to fund a cyclotron at the University of KwaZulu-Natal to generate useful variants of maize, millet and sorghum. It invested in an *in vitro* plant propagation platform, a study to assess the commercial viability of planting cassava in northern KZN to supply industrial starch to the South African paper-making industry, as well as projects to improve wine grape varieties and to protect cassava from Cassava Mosaic Disease.

The National Biotechnology Advisory Committee

The second recommendation made in the 2001 Biotechnology Strategy was the establishment of a National Biotechnology Advisory Committee (NBAC). It took rather a long time for this advice to be acted upon, firstly because the money for its implementation was made available only in 2003, and secondly because the Department of Science and Technology wanted all the BICs to get up and running without too many 'bosses', or committees, pulling them in different directions. The BICs were to become established with their own boards in place and comfortable with their governance, before the NBAC was created. The fear was that the BIC boards might defer decisions to the NBAC, which was not its purpose as it was meant to provide national strategic direction on biotechnology as a whole.

In November 2006, the NBAC was established as a sub-committee of the National Advisory Committee on Innovation (NACI). I was its first chair. Its mandate was to advise the Minister of Science and Technology on an appropriate course and suitable interventions for the development of biotechnology. Of the 10 original committee members, seven were still in office in 2012, which I think says a great deal about the commitment and enthusiasm of this group for their task.

Over the years we have given advice first to Minister Mangena, then to Minister Pandor and most recently, to Minister Derek Hanekom. I had the advantage of knowing Naledi Pandor when she was on the academic staff at the University of Cape Town, and also of having taught her daughter, Aisha. Towards the middle of Aisha's final year of BSc, her mother, then Minister of Education, came out with a statement to the effect that if undergraduates were failing they must be failed, not put through artificially. This would not be fair to the institutions, the students or their parents and could potentially lead to a drop in standards. At that time I was about to fail Aisha in her mid-year exams! I called her into my office, read her the riot act (after all her great-grandfather was the renowned academic ZK Matthews), and at the end of that year she passed with 70+ per cent. She actually wrote me a letter thanking me for giving her a wake-up call and she has now completed her PhD in human genetics. In 2010, at a meeting of Women in Science and Technology where Naledi Pandor was the keynote speaker, she told us that Aisha had asked her to babysit her child on a Sunday afternoon to allow her to go to the lab, as 'experiments don't respect weekends'. 'I have new respect for women,' she said, 'especially those with particularly small children, who persevere to complete their higher degrees, especially in experimental sciences.'

The NBAC has given the ministers advice on topics including funding for biotechnology, the regulation of stem cells, understanding the impact of public perceptions of biotechnology, and the development and retention of human capital for biotechnology. We have pointed out problems that were being encountered with decisions made by the executive council of the GMO Act (the most recent of these was mentioned in Chapter 2), and problems with issues of labelling of GMOs in the Consumer Protection Act of 2011. Some of these have been acted on, but many require consultation with other government departments, so only time will tell how successful these have been.

At its inception the members of the NBAC were mainly academics, but as our work progressed we were able to include more members from the private sector. This has proven invaluable as these are the people who will make a difference in bringing good ideas to the market place.

A calamitous development

The third recommendation from the 2001 Biotechnology Strategy was that government should articulate a single policy position on biotechnology. This has been a theme in many of our advice letters and is an issue because biotechnology is a cross-cutting discipline and needs close cooperation from many government departments. One of the problems with implementing the GMO Act is that the various departments involved—Science and Technology, Agriculture, Water Affairs and Forestry, Health, Trade and Industry, Environment and Tourism, and Labour—simply do not have a single policy position on biotechnology. Ignorance is sometimes the reason for this but when AfricaBio, the local version of the International Service for Acquisition of Agribiotech Applications (see Chapter 2), tried to run a training course for this very purpose, hardly any representatives came.

Government had been aware of the need for coherent action in the area of innovation. Indeed, in the National Research and Development Strategy of August 2002, they discussed the need for the establishment of a core agency to stimulate and intensify technological innovation. This eventually took shape, and on 24 November 2008 the Technical Innovation Agency (TIA) Act was passed by the President. The objective of the agency 'is to support the state in stimulating and intensifying technological innovation in order to improve economic growth and the quality of life of all South Africans by developing and exploiting technological innovations'.

By rights this should have been a great step forward for the biotechnological community, as the Department of Science and Technology's 10-year innovation plan (for 2008–2018) stated that 'over the next decade South Africa must become a world leader in biotechnology and the pharmaceuticals'. But has this been the case? The department was not unhappy with the functioning of the BICs, although there was obviously room for improvement. They wanted a more national approach and a more standardised *modus operandi*. They had deliberately allowed the BICs to pursue independent approaches with the intention that these would be reviewed over time and lessons learned to aid greater standardisation. However, with the advent of the TIA, the BICs were to be dissolved and incorporated within the new structure. The idea had been that their experience, systems and expertise would help to develop and strengthen biotechnology within the new organisation. That was the idea, but in practice the disruption to the biotechnology community has been little short of disastrous.

As early on as 2008, the NBAC had begun to feel very uncomfortable about the form the TIA was taking and the functions it was appropriating. Hence an advice letter was presented to the minister at a meeting on 15 January 2009. It stated as follows:

(1) It is recommended that under the TIA the skills (expertise) in, focus on, and funding of biotechnology, currently represented by the BICs, must be maintained.

(2) It is recommended that the budget allocated to biotechnology within the new TIA structure be increased significantly.

(3) It is recommended that funding security is guaranteed over the medium term (at least 10 years duration) for biotechnology.

(4) It is recommended that a formal and functional link be established between NBAC and the TIA board.

The notes on the meeting read as follows:

> *Minister Mangena responded that the advice was timely since TIA was in the process of being established. He mentioned that concerns similar to that of NBAC had been raised by all other entities due to be incorporated into TIA. The new TIA Board would be representative of all the entities entering TIA. Advice on the most optimal structure for TIA would be useful. The Minister assured the meeting that biotechnology is regarded as a very important field and that the Department of Science and Technology would make sure that the focus on biotechnology would not be lost because of TIA, since the reason behind the establishment of TIA was to move forward in this and other fields of innovation.*

The TIA board was established in July 2009 under the chair of Dr Mamphela Ramphele, former Vice-Chancellor of UCT and Deputy President of the World Bank. In September 2010 the NBAC held a national workshop entitled 'Feeding the biotechnology pipeline'. Participants included researchers, academics, workers in international property, and representatives from industry. The aim of the workshop was to highlight gaps in the pipeline from the initial idea to its commercialisation in the open market. Although many gaps were discussed, the major problem identified by almost all participants was the TIA, with one of the issues being the sudden demise of the BICs. Some scientists more closely involved with these four organisations might have been aware that they were to be merged into the new structure, but even they were amazed when the BICs were summarily closed. Indeed, members of my lab had submitted a grant proposal to PlantBio early in 2010, only to receive a short letter towards the end of March to say that they would no longer be in existence after the end of the month and to please refer all queries to the TIA. No contact details were given and no letters or emails to the TIA were replied to, nor were phone calls returned.

The knock-on effects of this to the biotechnology community were expressed extremely strongly at the workshop: scientists were being retrenched and projects shut down. Moreover the emphasis was only on short-term, low-risk return on investment, whereas in a study on the time it takes and the costs involved in getting a good idea to the market when the product involves a GM crop, these factors were estimated as 13 years and US$136 million (see Chapter 4). One participant at the NBAC workshop went so far as to say that the biotechnology community in South Africa was haemorrhaging scientists.

On the same subject the Vice-Rector (Research) of the University of Stellenbosch, Prof Arnold van Zyl, wrote a commentary on the role of the TIA in the *South African Journal of Science* (Van Zyl, 2011). He noted that the agency, whose object it is to 'support the state in stimulating and intensifying technological innovation' had not apparently done so. 'On the contrary, quite a number of funding initiatives incorporated into the TIA have been abruptly ended, leaving research institutions responsible for personnel and running costs, and in some cases even resulting in the loss of highly skilled personnel.'

Shortly after the workshop, NBAC asked to meet with the CEO of the TIA, Simphiwe Duma, to discuss these problems, but to no avail. Continued requests were met with silence until eventually we were told that he would meet with us on 15 July 2011. I arrived in good time from Cape Town, only to be told half an hour before the meeting that he was unable to attend. I expressed my extreme displeasure and finally, after the other NBAC members had gathered, two members of the TIA arrived. The meeting was not particularly helpful as they were reluctant to concede that there were any problems. They did, however, acknowledge that 'the process of migrating the BICS to TIA took longer than was anticipated and *might have inconvenienced some stakeholders within the system*' (my italics).

In September 2011, the Agricultural Biotechnology International Conference was held in Johannesburg with the title 'Moving towards a bioeconomy—agricultural biotechnology for economic development'. I was chair of the scientific organising committee, and was extremely relieved when we secured a senior official in the TIA, Dr Bongi Gumede, to give a presentation on his organisation's involvement in biotechnology. When I greeted him just before his session, to my horror he told me me that instead of him, Caiphus Ramoroka, who had been in office for precisely six days, would present the talk. The reason Dr Gumede gave was that in future the agricultural community would be dealing with Mr Ramoroka. When I remonstrated with him, saying we wanted to hear from someone well versed in the recent turbulent relationship between the TIA, NBAC and the biotechnology community in general, I was simply told that this would not occur. As I anticipated, we were given an extremely general overview of the mandate of the TIA with no reference at all to the problems being encountered by biotechnologists in South Africa. In March 2013 the new Minister of Science and Technology, Derek Hanekom, instituted a panel to review TIA. The Bioeconomy Strategy Group of NACI was invited to give an oral presentation on Friday 8 March, but our meeting was cancelled by the panel. We still hope that our voices will be heard.

One of the most successful spin-offs of PlantBio, which has fortunately survived within the TIA (although for how long is at present unknown), is Biosafety South Africa. Under the able direction of Dr Hennie Groenewald it is involved in helping to ensure that regulators and technology developers have the necessary information and know-how on compliance matters. In 2011 he attended the annual general meeting of an organisation called SACAU—the Southern African Confederation of Agriculture Unions. He said it was one of the most stimulating meetings he had attended in a long time as he was interacting with people who really knew, from their own farming experience, what they were talking about. After the meeting, SACAU adopted a GMO policy framework based on their acknowledgement that 'GM technology is one of the options that can increase production, improve productivity and incomes of farmers, and contribute to addressing food security challenges in the region'. They recognised the following points:

- The need for evidence-based decision making
- The importance of directly involving farmers in R&D and related standards
- The right of consumers to choose
- Cost/benefit analyses should also look at the costs associated with non-adoption of GMOs
- The need to monitor trade in the region
- The importance of political will and harmonisation of policies in the region (SACAU, 2013)

It is the collective voices of farmers' organisations such as SACAU that will be imperative if GM crops are ever to be able to help bring about food security in Africa.

The Bioeconomy Strategy

Early in 2013 the Department of Science and Technology (DST) approved a new Bioeconomy Strategy to replace the old 2001

Biotechnology Strategy. 'Bioeconomy' refers to 'activities that make use of bio-innovations, based on biological sources, materials and processes, to generate sustainable economic, social and environmental development'.

Under this umbrella the entire innovation system or network, from the initial ideas, research and development through to production, manufacturing and commercialisation, are coordinated. The main problem seen with the original strategy was that it focused on harvesting opportunities that represented a quick return on investment. It supported commercialisation of technologies close to the market. The present strategy is formulated to develop the entire value chain for biotechnology-based products.

The DST correctly sees itself as the key driver in this initiative and undertakes to coordinate the stakeholders and role players. As one of the most serious flaws in the implementation of the GMO Act has been the lack of coordination between the various government departments involved, it is hoped that this time around these problems will be overcome. The document also states, under the section on agriculture, that genetic engineering remains a critical technology and presents a significant competitive opportunity for sector development. As it often appears in the administration of the GMO Act that some government departments do not share this opinion, it will be interesting to see whether the DST is able to convince them otherwise.

The document compares South Africa's performance in the bioeconomy with countries including Australia, Brazil, Cuba, India, Malaysia and Singapore. From this it is clear that we need to increase significantly our gross expenditure on relevant research and development. It is important, however, that such funding should capitalise on our opportunities and strengths such as the fact that we are the third most biologically diverse country in the world.

At the end of the document, under 'The way forward', it states:

> *The Department of Science and Technology, as the lead agent of this strategy, will continue to engage with line departments to promote cooperation, facilitate the strategy's broad implementation, and ensure synergy, alignment and better coordination of activities. The Department of Science and Technology will drive a consultative process to define the roles and responsibilities of various government departments, agencies and instruments in implementing the strategy* (The Department of Science and Technology Bioeconomy Strategy [version 9], 11 February 2013).

All strength to their collective aim.

Chapter 6

African National Biotechnology Strategies

In 2001 the United Nations Environment Programme (UNEP) and the Global Environment Facility (GEF) set up a Global Project for the Development of National Biosafety Frameworks. One of their aims was to help more than 100 countries worldwide to set up frameworks that would enable them to undertake work using GM crops. When this project ended in 2006, 23 African countries had completed their frameworks. To date about 40 have been finalised. As Robert Paarlberg explains (Paarlberg, 2012), most of these countries embraced the strongest possible approach, the 'Level One' approach, following European governments, which requires new legislation. Most countries cited the lack of a suitable existing law that could serve as a 'home' for using this approach. Fewer followed the approach taken by the US, which uses existing laws to regulate GMOs, the so-called 'Level Three' approach. In the three other key requirements for the use of GM crops, African countries have also followed Europe rather than America. The other three requirements are:

(1) the creation of new institutions such as national biosafety committees to oversee GM crop management
(2) the use of the 'precautionary approach', which allows regulators to decline to approve a new technology on the grounds of 'uncertainty', without any evidence of risk
(3) the requirement for labelling products derived from GMOs in the marketplace.

In many ways, Europe's attitude to GM crops is not based on risks, but rather on the absence of any new benefits. The first generation of these crops—insect- and herbicide-resistant maize, cotton and soya bean—benefited farmers, rather than consumers. Moreover, Europe does not have many maize, cotton and soya bean farmers, so the technology had few champions. But this is not the case in Africa, where farmers are often consumers and maize, in particular, is often the staple diet. So why does Africa follow Europe rather than America?

Foreign aid is the greatest external influence in Africa, and Europe gives three times more aid than the US. European governments also give roughly three times as much as America does to the Global Environment Facility, and the international NGOs active against GMOs are mainly based in Europe. A fourth channel of external influence is cultural. Most policy-making elites in Africa have much closer cultural ties to Europe than to the US, and are thus more likely to follow the European anti-GMO stand. Finally there is the question of international trade. Africa's farming community's exports to Europe are six times greater than those to the US, and so European consumers call the shots. In fact, South African exports have not been affected by its plantings of GM crops, but these anxieties influence the thinking in many African countries (Paarlberg, 2012).

I personally saw the effect of international NGOs at the 2002 World Summit on Sustainable Development in Johannesburg when European groups instigated a 'farmers' march (with little evidence of farmers in the foreground) saying 'NO to GM foods'. In addition, Friends of the Earth encouraged their African partners to sign an open letter warning that GMOs might cause allergies, chronic toxic effects and cancers (Paarlberg, 2012).

In a hard-hitting article, Giddings et al. (2012) argue that the Gordian knot binding European plant science through continuing policy failure and political timidity will remain uncut unless bold action is taken. They quote Jared Diamond, who wrote in *Guns, Germs and Steel* (1998:257):

> *Any society goes through social movements or fads, in which economically useless things become valued or useful things devalued temporarily. Nowadays, when almost all societies on Earth are connected to each other, we cannot imagine a fad's going so far that an important technology would actually be discarded. A society that temporarily turned against a powerful technology would continue to see it being used by neighbouring societies and would have the opportunity to re-aquire it by diffusion (or would be conquered by neighbors if it failed to do so).*

Giddings et al. (2012) go on to say that the world has seldom seen a greater discrepancy between the inherent hazard of a product and the level of regulatory burden imposed on it than exists today for crops improved by biotechnology. 'It is important, here, to be very clear: there is no basis *in science* (my emphasis) for regulation specific to crops and foods improved through biotech or GMOs.' Indeed, by any honest reckoning, 'the level of scrutiny to which crops improved through biotech are subjected is completely unwarranted by the body of knowledge acquired over three decades of experience with such crops, including 15 years in commercial production'.

They state further, quoting from Brookes and Barfoot (2011), that since 1996, farm incomes have increased by US$64.7 billion, and in 2009, 53.1 per cent of the farm income benefits have been earned by developing country farmers. Moreover, between 1996 and 2009, the cumulative farm income gain derived by developing country farmers was 49.2 per cent, equating to US$31.85 billion. In terms of environmental impact, since 1996 the use of pesticides on biotech crops was reduced by 8.7 per cent, and the environmental impact associated with herbicide and insecticide use on these crops, as measured by the EIQ (environmental impact quotient) indicator, has fallen by 17.1 per cent.

Despite this, Europe is still caught in its Gordian knot: 'Nowhere is the chasm between regulatory regimes and the implications of facts and experience greater than in Europe.' But this is the model which African countries are following.

Paarlberg identifies three critical factors that must be in place for a GM crop to be approved:

(1) There must be a functional national biosafety committee, legally constituted in a separate act of parliament, according to the 'Level One' approach mentioned above.
(2) The GM crop must have demonstrated strong agronomic performance, outweighing anything currently available using conventional technologies.
(3) There must be strong in-country political support, sufficient to overcome the organised opposition.

The development of national biosafety strategies has taken African countries many years. Charles Gbedemah was running the UNEP/ GEF programme from his office in the United Nations compound in Nairobi between 2001 and 2007, and mid-way during this exercise I visited him. I discovered that 132 countries were involved for a variety of reasons that included both national development priorities and international obligations—for instance, to the Cartagena Protocol on Biodiversity. For some countries, however, the primary reason for joining the project was to have access to funds from UNEP/ GEF for capacity-building activities. I cynically wondered if these capacity-building activities did not include funding for personal travels to and from Nairobi and other international centres for 'training purposes', leading to a lack of urgency to terminate the process.

The following six African countries have used slightly different approaches to arrive at a National Biosafety Strategy.

Kenya

The first GM crop to be considered for adoption was *Bt* maize. Kenya's national agricultural research institute (KARI) started confined field trials in 1999, but the crop didn't get any further. In this case all three of the above conditions were missing. There was no legislation, the crop controlled only the stalk borer pests in part of the country and, partly as a result, there were few champions (Paarlberg, 2012).

In 2009 the Biosafety Act (Act No. 2 of 2009) was passed. Margaret Karembu and colleagues at the ISAAA *Afri*Center in Nairobi have written an instructive account of the lessons learned from this experience (Karembu et al., 2010). It underlines the importance of Paarlberg's point number 3: strong in-country political support. Although the first draft of the bill was ready by the end of 2002 it was stopped due to a general election. The new government had lost many of the MPs who had been sensitised to the importance of biotechnology, so the pro-biotech stakeholders had to acquaint themselves with the new MPs in order to cultivate champions among them. The year 2004 was marked by the clamour for a new constitution, and MPs had different issues on their agendas. A silver lining was the creation of a Ministry of Science and Technology and the appointment of Dr Noah Wekesa, a scientist, to head it. The bill was finally published in the *Kenya Gazette* in 2007, but then widespread violence followed the presidential election.

I experienced some of the effects of this violence, albeit second-hand. I was about to start a new PhD student from Kenya in January 2008 and needed documents from her home institution. She managed to get a message through to me from her rural home village saying it was not safe for her to travel to Nairobi, and asking whether she could delay her arrival. That went without saying, but when she started at UCT some months later it was clear that the experiences had been traumatic.

After the violence subsided, the biotechnology stakeholders formed themselves into the Biosafety Consortium and two new ministers came to their aid: Dr Sally Kosgei, from Higher Education, Science and Technology, and William Ruto in Agriculture. There was a flurry of awareness-creation activities targeting MPs and, in due course, the bill was passed and signed by President Kibaki on 12 February 2009, ending a journey of almost 10 years.

However, before the Act could be implemented, in order to allow the commercialisation of GM crops regulations had to be passed. This was delayed by the fact that, in November 2010, William Ruto was identified by the International Court in the Hague as among six Kenyans to be investigated for crimes against humanity emanating from the post-election violence in early 2008. (Ruto was elected Vice-President of Kenya on 9 March 2013 together with President Uhuru Kenyatta, a co-accused at The Hague.) However, the regulations were passed by Sally Kosgei in July 2010. Kenya is therefore poised to become the fourth country in Africa, after South Africa, Egypt and Burkina Faso, to commercialise GM crops. However, at the point of going to press, this had yet to happen.

Uganda

Uganda has had a GMO Act in front of parliament since 2005 but it has never been passed. There are a number of reasons for this. Firstly, although it cannot commercialise GM crops, it can approve glasshouse and field trials. This is done by the Ugandan National Council for Science and Technology and they have already approved trials for banana, cotton, maize and sweet potato. Uganda has therefore put itself into a contented 'research forever' situation whereby it keeps the grant money coming in, its scientists content and the opposition happy.

Who is the opposition? Organic farmers are a strong lobbying force in the country. According to Robert Paarlberg (personal communication), the National Organic Movement of Uganda claims to have 250 different member and partner organisations. Uganda is indeed an attractive country for promoters of organic farming because most farmers are too poor to use any synthetic fertilisers, making them *de facto* organic and hence easy to certify. This movement is not only against GM crops on principle (although I have often wondered how a quintessentially organic molecule such as DNA can be classified as 'non-organic'), but also out of commercial fears that European importers would doubt the organic status of products planted in a country which allows GM crops to be planted.

Another reason for the Act sitting dormant is the lack of a support group pushing for its passage through parliament. Unlike neighbouring Kenya, where farmers joined together to form an effective lobbying group to encourage the government to pass the Biosafety Act, in Uganda very few farmers plant hybrid crops such as maize. As GM crops are all hybrids, what is the urgency to use these? On the contrary, the Ugandan climate and environment is such that in recent years it has been a maize-exporting country, mainly to other African countries. In fact, until 1992 there was no national farmers' organisation, and certainly not one to push for the adoption of GM crops such as maize.

Finally, by some strange institutional quirk, the ministry responsible for the bill is not the Ministry of Agriculture, or of Science and Technology, but of Finance, Planning and Economic Development. This ministry is an extremely important one and the GMO Act is not high on its agenda (Paarlberg, personal communication).

It therefore appears that none of the three requirements outlined above, which are critical to ensure some level of urgency for the commercialisation of GM crops, is in place. It may transpire, therefore, that the GMO Act will continue to sit dormant in the Ugandan parliament for the foreseeable future. That might change, however, if a new GM crop which has singular benefits for farmers appears on the horizon—and that might well be in the form of pest-resistant bananas, as discussed in Chapter 4, and the bacterial wilt-resistant bananas managed by the AATF, as I will outline in Chapter 9.

Tanzania

Tanzania is one of the few African countries to follow, albeit probably not intentionally, the lines of the US in not having a separate national biosafety law for GMOs, but instead using a multitude of existing laws, guidelines and regulations. Tanzania is also different from Kenya and Uganda in not having approved any field trials for GM crops, and I cannot envisage this situation changing in the near future.

The main reason for this is their 'strict liability' system for responding to any alleged social damages following the introduction of a GM crop. This could put technology providers such as multinational seed companies at a prohibitive risk. Most countries have a 'fault-based' system for finding technology providers liable for damage. This requires evidence of bad faith, negligence, and a cause-and-effect link between an action and the damage (Robert Paarlberg, personal communication). The 'strict liability' precaution is usually reserved for companies handling inherently hazardous substances or technologies, such as toxic waste or nuclear power. However, under Tanzania's 'strict liability' requirement, if a drought-tolerant GM crop failed due to a hurricane or an invasion of locusts, the seed company who sold the farmer the crop could be liable for damages. It would be just too easy for GMO critics to claim damages and initiate legal action.

Even GMO promoters in Tanzania, such as scientists, are happy with the status quo, as most of them are focused on permission to begin research, not to apply for environmental release. This could have a bearing on the testing of drought-tolerant crops, but at the moment it appears that no field tests of drought-tolerant maize will be carried out in Tanzania.

Burkina Faso

Burkina Faso surprised many of us by suddenly allowing the commercialisation of *Bt* cotton in 2008. One of the reasons for our surprise was that they did not even have a Biosafety Act in place when they approved this. However, the country's legislation had passed biosafety legislation to formalise regulatory oversight for both the research into, and the commercialisation of, agricultural biotechnology products (Vitale et al., 2011). This biosafety law 'pertaining to the security system as regards to biotechnology in Burkina Faso' was passed in 2006, establishing Burkina Faso as the first West African country to have enacted such a law. The bill established the National Biosafety Agency (NBA) as the regulatory body, with members of several government agencies and NGOs. Although not established by the law, the NBA is led by and housed in the Ministry of Environment (Birner et al., 2007).

From 2003 to 2005, the Institut National de l'Environnement et Recherches Agricoles conducted confined field trials that evaluated the effectiveness of the crop within the climate and insect conditions specific to the country. In 2006 the testing was more widespread and the approval came in June 2008 (Vitale et al., 2011). In 2009, following a decade of coordinated efforts on behalf of various cotton stakeholders to satisfy a series of technical, legal and business requirements laid down by the legislature, farmers planted about 125 000 hectares of insect-resistant cotton.

Based on a survey of 160 cotton producers, Vitale et al. found that:

- *Bt* cotton increased cotton yields by an average of 18.2 per cent
- there was no difference in production costs since the increased cost of seed was offset by a reduction in insecticide costs
- *Bt* cotton producers earned a profit of US$ 39.00 per hectare, which shifted producers' bottom line from a negative position to a positive one.

The authors noted that part of this improvement might have been due to the fact that only three cotton companies were connected to the vast network of smallholder producers. They compared this to a similar situation reported by Gouse at al. (2003) for South Africa where a single cotton company provided inputs to producers and was the sole buyer of cotton. This immediately raises alarm bells with me, as the South African *Bt* cotton market is currently in a downturn, precisely because it was dependent on one company which experienced financial difficulties, due partly to mismanagement and partly to the uncompetitive nature of the cotton crop (*Bt* or not) in that region. But the involvement of three companies in Burkino Faso will, it is hoped, prevent such an unfortunate occurrence, which has nothing to do with the GM technology itself.

Reports from plantings in Burkina Faso in 2012 indicate that somewhat more than half (51 per cent) of the total 615 795 hectares planted to cotton was *Bt*, involving approximately 100 000 farmers (James, 2012). The country is also going ahead with field trials of *Bt* cowpea.

Of some concern is a proposed amendment of the Biosafety Law which was recently drafted to give more political autonomy to the NBA, but also to provide some strict liability provisions. The strict interpretation of these could make it prohibitively costly to deliver GM crops to farmers (IFPRI, 2012).

Ghana

Ghana's Biosafety Act was passed into law towards the end of 2011. This was the culmination of years of outreach activities and advocacy for biotechnology, involving a cross-section of Ghanaians, including the scientific community, the media, farmer-based organisations, consumer associations, religious bodies and legislators supported by donor agents or their representatives (Walter Alhassan, personal communication). Speaking at a public meeting in Accra in early 2012 on behalf of the Forum for Agricultural Research in Africa (FARA), Professor Alhassan said that 'having the law means the work is now beginning' and allows for the production of biotech foods on a commercial scale.

Egypt

Of the 23 countries that had completed their National Biosafety Frameworks under the United Nations Environment Programme by 2006, only the relative newcomer, Burkina Faso, was listed in the top 28 countries in the world planting GM crops in 2012 (James, 2012). It comes in at number 14 (0.3 million hectares of cotton). South Africa is number eight, with 2.3 million hectares planted to GM maize, cotton and soya beans. Egypt, which began planting *Bt* maize in 2008, and has the well-established Agricultural Genetic Engineering Research Institute (AGERI), comes in at only number 25 (<0.1 million hectares). In 2008 they approved the cultivation and commercialisation of *Bt* maize, although this is currently under review.

AGERI is very active in the development of GM crops. They are working on insect- and virus-resistant potatoes, virus-resistant tomatoes and cucurbits, and drought-tolerant wheat. Unsurprisingly they also have a keen interest in improving their cotton varieties. They are in partnership with Monsanto to cross Egyptian elite germplasm with Bollgard II (one of Monsanto's *Bt* insect-resistant cotton varieties) to form Giza-Bollgard II. They aim to pave the way for acceptance of GM cotton in neighbouring African and Asian countries.

• •

With so few countries in Africa passing biosafety laws, it would seem that Europe is winning its war against the acceptance of GM crops on this continent. Although many African countries had completed their national biosafety frameworks under the auspices of the United Nations programme by 2007, only four are able to commercialise such crops. I wonder if the Europeans behind this approach would be proud of themselves if they were to visit some of the women and men in Africa desperate to use this technology to elevate themselves out of the status of subsistence farmers to that of commercial farmers. A meeting of stakeholders involved in agricultural biotechnology was held in February 2012 in Nairobi under the auspices of the African Development Bank. If the Bank were to come out in favour of GM crops this could have an enormously positive effect on African countries' future adoption of this technology.

Chapter 7

The maize streak virus story

The title of our research group's paper in the *Plant Biotechnology Journal* read 'Maize streak virus-resistant transgenic maize: a first for Africa' (Shepherd et al., 2007) with a photograph of an infected maize cob on the front cover (Figure 7.1). The journal *Science* led with 'GM technology develops in the developing world' (Figure 7.2; Sinha, 2007). Success at last, but when we began this project some 20 years before, I was woefully ignorant of what it would entail, and if I'd known then what I know now I would probably never have embarked on the journey. For a start I didn't realise (and I really should have) that you can't grow maize in the Cape Peninsula—well, not properly, so that it can be healthy, vigorous and set seed. And you can't grow GM maize unless you have a computer-controlled glasshouse, and at the time we didn't even have a glasshouse at all. Instead, we grew our maize in securely locked, artificially lit, growth chambers in the basement of our building on the UCT campus.

Maize streak disease (MSD) was first described in 1901 when it was found on plants in South Africa in what was then Natal, as follows: 'The disorder of the mealie plant, locally described as "Mealie Blight", "Mealie Yellows" or "Striped Leaf Disease", belongs to a group of plant troubles arising from obscure causes ...' (Fuller, 1901). Fuller mistakenly attributed the disease to a soil disorder, but in retrospect it is quite clear that the 'mealie variegation' he described and drew in minute detail can be attributed to maize streak virus (MSV, Figure 7.3). In the 112 years since this first report, scientists have come a long way in identifying and analysing the causal agent of MSD, to the point where in our lab we believed we could design effective strategies to control, or even eliminate, the disease in maize.

When I joined UCT as the Head of the Microbiology Department in 1988, I inherited a department divided between the second floor

Figure 7.1   Cover of *Plant Biotechnology Journal*

AGBIOTECH

GM Technology Develops in the Developing World

The first genetically modified crop developed entirely in Africa is gearing up for field trials. Its success would be a milestone

Unscathed. Unmodified plants (*left*) show signs of maize streak infection, but the GM plants (*right*) are symptom-free.

Figure 7.2   Headline from *Science* with photo of transgenic maize streak virus-resistant and non-transgenic maize

bacterial geneticists and the third-floor plant virologists. How to merge these two rather headstrong groups? Ever since I had met Marc van Montagu, one of the trio responsible for the development of plant genetic engineering, during the European Molecular Biology Organization course in Basel in 1979, and spent time in his lab in Ghent, I had wanted to move into this field of research. When running the CSIR's Laboratory for Molecular and Cell Biology, I had worked on the bacterium, *Agrobacterium tumefaciens*, the organism Marc and his colleagues had developed as a vector for introducing foreign genes into plants. So, although I was familiar with working with this bacterium, I had not progressed to using it to genetically modify plants. But now I saw my opportunity to do this and I hoped to bring the two divisions in the department together. Ed Rybicki, then senior lecturer and now professor, was and still is South Africa's leading plant virologist. He and I put our heads together to combine our expertise to develop maize resistant to the African endemic maize streak virus (MSV).

Figure 7.3 Symptoms caused by MSV on a maize plant

Why were Ed and I so passionate about this project? MSV is the most significant pathogen of maize in Africa, resulting in crop yield losses of up to 100 per cent. Transmitted by tiny leafhopper insects,

Cicadulina mbila, it is indigenous to Africa and neighbouring Indian Ocean islands. Despite maize being a crucial staple food crop in Africa, the average maize yield per hectare on the continent is the lowest in the world, leading to food shortages and famine. A major contributing factor to these low yields is MSV (Figure 7.4). Indeed, I have often said to small-scale farmers whose fields I have visited in Kenya, 'Why do you bother to plant maize? It will just be destroyed by the virus.' Their reply is that 'Any cob is better than no cob'.

Figure 7.4   Smallholder farmer in a field of maize infected with MSV

Thus our aim was to develop maize resistant to MSV. However, we knew that we had to cut our teeth on something simpler as, in those days, nobody had succeeded in using *Agrobacterium tumefaciens* to genetically engineer monocots, such as maize, but only dicots, such as tobacco. *A. tumefaciens* is a soil bacterium which can infect dicot plants and, as the name implies, cause a tumour. Marc van Montagu and his colleagues discovered that when the bacterium infects a plant it transfers part of its DNA into the plant cells where it becomes integrated into the plant's genetic material. Not long after this discovery they realised that the introduction of a foreign gene into *A. tumefaciens* DNA would enable its transfer to the plant cell nucleus. Moreover, they could remove the genes that caused the tumour without affecting the transfer of this genetic material. This is what led to the development of dicot plant transformation.

Ed and I accordingly devised a project to be carried out by a willing and very able PhD student, Andy Hackland. Most plant viruses have, as their genetic material, RNA, a structurally simpler form of the normal genetic material, DNA. One of the genes on this RNA molecule codes for the virus coat protein (CP). When a virus infects a plant, it injects its RNA, leaving its CP outside. This RNA replicates and codes for the proteins required by the virus to infect the plant. What scientists had shown was that if they made transgenic plants carrying the CP gene, the viruses would not replicate, and therefore no infection would occur. The simplest way to think of this mechanism is that, as soon as the virus RNA enters the plant cell, the CP produced by the plant will recoat the RNA and no replication will occur. This is an over-simplification of the mechanism but serves to illustrate the point.

The project was therefore to clone the CP genes from cauliflower mosaic virus (CaMV) and tobacco necrosis virus (TNV) and introduce them into tobacco using *A. tumefaciens*. The CaMV CP had been shown by others to produce resistance, so we used it as a control. TNV-resistant tobacco would be a first and, as this virus causes disease in tobacco, it could have commercial interest. However, that was not our main aim, which was simply to show that we could carry out this research with the people, equipment and facilities we had at UCT. Needless to say, the project worked beautifully and Andy duly graduated with a PhD in 1993. He then went on to start a plant tissue culture company, Frontier Laboratories.

Help from abroad

The next step on our journey was to learn how to transform maize, a monocot which, at the time, was resistant to *A. tumefaciens* transformation. Bill Gordon-Kamm, then working for the American biotechnology company, DeKalb, in Connecticut had published a seminal paper on this subject in July 1990, entitled 'Transformation of maize cells and regeneration of fertile transgenic plants' (Gordon-Kamm et al., 1990). I contacted him and asked if I could visit. Bill couldn't have been more helpful, and when I told him that we wanted to develop maize resistant to a devastating African endemic virus he was most enthusiastic, with an obvious desire to help developing countries. I therefore arranged that my tissue culture technical assistant, Sandy Lennox, would spend some time in his lab and learn the technique.

Next I visited Ciba-Geigy, whose Ciba Agricultural Biotechnology division was based in the Research Triangle Park in North Carolina. After a number of mergers, they are now part of Syngenta. Mary-Dell Chilton, one of the trio who developed *A. tumefaciens*-based plant transformation, was the director, and she kindly facilitated my visit. The only fly in the ointment emerged when her staff took me round their glasshouses. These were mind-blowingly sophisticated—and they told me in no uncertain terms that I shouldn't even contemplate going into maize transformation if I didn't have a computer-controlled glasshouse. This, they proudly told me, allowed them to set maize seed almost 12 months of the year. I couldn't bring myself to tell them that at home I didn't even have access to a glasshouse!

Back at UCT, what to do about this rather significant shortcoming? At the time I served as a senate representative on the University Council and so I hesitantly approached the chair with my problem. He must have moved a minor mountain because within a year UCT had a brand new glasshouse to which I had access. It was not computer-controlled and, with UCT situated in the lee of Devil's Peak, lost the sun fairly early, especially in winter, but at least it was a glasshouse.

The next problem to be solved was funding. Although I received some limited funds from the government-backed National Research Foundation, I knew it wouldn't be sufficient to embark on this ambitious project. The Claude Leon Foundation (known in those days as the Claude Harris Leon (CHL) Foundation) had generously funded the Departments of Microbiology and Biochemistry for a number of years, enabling them to buy expensive state-of-the-art equipment. However, when I joined UCT they were indicating that this funding would probably end shortly as, over a 10-year period, they had supported the departments to the tune of over R1 million. I was extremely grateful for the assistance they had given the Microbiology Department under my predecessor Prof David Woods, and considered the CHL Foundation had every reason to spread their largesse. And that would have been the end of it if it hadn't been for the intervention of our then Vice-Chancellor, Stuart Saunders. He told me that if I applied to the foundation for funding of my research programme specifically, instead of general departmental funding, I might find a receptive ear. So I did, and thus began six years of a most fruitful and beneficial relationship, which I cherish and am immensely grateful for to this day.

The Claude Leon Foundation

Claude Leon had made his money as founder and managing director of the Elephant Trading Company, a wholesale business based in Johannesburg. He also helped to develop several well-known South African companies, including Edgars, OK Bazaars and the mining house Anglo Transvaal (later Anglovaal). The trustees of the foundation, mainly UK-based grandsons of Claude Leon, visited South Africa once a year to check up on the projects they were funding. One trustee, Brian Yule, was extremely diligent in checking up on my lab's financial situation and required me to present a breakdown of how the foundation's funding fitted in with the overall funding of research in the department—quite a tall order for someone like me, for whom finance is not a strong suit.

On one occasion, Brian was joined by his wife, Annie, a molecular biologist. She was almost as stern as Brian and, during our discussions, asked me how many post-doctoral fellows I had in my group. When I said, 'None', she was clearly horrified and asked me how on earth I managed to run a research team without them. I explained that there was not a 'culture' of post-docs in South Africa, as newly graduated PhD students either went abroad to do such a post-doc, or looked for a job. This turned out to be a seminal discussion for South African science, as the next project the foundation funded was the awarding of post-doctoral fellowships, which by 2012 (the 15th year of the programme) had reached a total of approximately 450. They asked the Royal Society of South Africa to help them with the evaluations and, for a number of years, I served as a member of the panel that undertook this task.

Learning the technique

With funding secured for our MSV project, the next step was to send Sandy Lennox to DeKalb and then to Ciba-Geigy to learn how to transform maize. The method used to transform maize in those early days is called particle bombardment or biolistics, a word derived from the words 'biological' and 'ballistics'. The name is derived from the fact that the early prototypes of the apparatus did, indeed, resemble a gun and the genes were propelled into the plant by gunpowder. The apparatus, although now powered by helium gas and looking very ungun-like, is still called a 'gene gun'.

To transform plants biolistically, DNA carrying the genes to be transformed into the plant is coated onto tiny, chemically inert metal particles, usually gold or tungsten. The gene gun is used to shoot these particles into the plant cells. A number of instruments are available for use in biolistics, based on various accelerating mechanisms. Figure 7.5 shows the most widely used gene gun, currently marketed by Bio-Rad, Inc (Biolistics®). After 'bombardment' the plant cells are regenerated into plants, due to the fact that plant cells are totipotent. This means that a single, isolated plant cell can grow into an entirely new plant. Thus if a gene is transferred into an isolated plant cell, every cell of the regenerated plant will contain this gene.

Not knowing anything about this technology apart from what she had read in Bill's paper, Sandy set off for Mystic, Connecticut towards the end of October 1993. Unbeknown to her, Bill had unexpectedly resigned while she was in the US on her way to his lab. But, faced with a *fait accompli*, the maize biotechnology research group at DeKalb welcomed her most professionally and gave her an excellent introduction to the technology.

Sandy went on to Mary-Dell Chilton's group at Ciba-Geigy in North Carolina. The staff in charge of the maize transformation work taught her how to excise immature embryos from young maize cobs and initiate the development of callus tissue. A painstaking process

Figure 7.5   The Biolistic® Accell® gene gun

then followed whereby the appropriate callus is excised out under precision microscopy for transfer into the gene gun for bombardment of DNA-coated gold particles. Thereafter the transgenic callus had to be selected on media containing the herbicide bialaphos which selects for those that are transgenic. As bialaphos is very expensive, we were extremely grateful that both DeKalb and Ciba-Geigy donated samples for our use back at UCT. Following regeneration, the developing plantlets have to be hardened off under specific growth chamber conditions which lead to maize growth, maturation, pollination and, finally, the setting of seeds.

Getting the equipment

Sandy returned after four weeks full of ideas and enthusiasm, but first we had to acquire the necessary equipment. This included the particle gun, laminar flow hoods to ensure a sterile working environment, dissection microscopes (including one with a camera attachment) and suitable shelving and lights for the plant growth rooms. Sandy now set to work to ensure we could repeat the processes she had learned in America, and it soon became apparent that all was fine except for getting our regenerated maize plants to set seed. The light intensity in the growth rooms (and even in the new glasshouse) was simply not high enough for this to take place. As I said earlier, you can't grow maize properly on the Cape Peninsula. We needed help—and into this breach stepped the company PANNAR Seed (Pty) Ltd.

PANNAR was founded in 1958 in Greytown, a prosperous agricultural community in KwaZulu-Natal, where its head office remains to this day. It is a privately owned, independent group of companies that is a regional market leader, being the largest seed group in Africa, as well as a global player. It owns companies in nine African countries, Argentina, Europe and the US, and markets seed across the globe. We couldn't have found a better partner.

It is no exaggeration to say that, had it not been for PANNAR, this project would not have succeeded. From a small beginning of R22 500 in 1995, the funding is currently somewhat more than R5 million. But it was not just money that PANNAR brought to the table. They helped us immeasurably in kind—from setting seed, making crosses, doing glasshouse trials, and giving advice to offering boundless enthusiasm as well as healthy criticism.

While all this was going on, a new student, Tichaona Mangwende, the first of three Zimbabwean PhD students from the MSc course I examined there in 1992, joined the group. He was brave enough to take on the task of MSV resistance, and the first question we had to answer was what gene to use. The tried and trusted approach used for most plant viruses was to transfer the gene coding for the viral coat protein into the plant. However, there was evidence that over-expressing the MSV coat protein would not work, so we decided to use the gene coding for the replication associated protein (Rep). MSV has a very simple DNA genome, with only three genes. One of them is *rep* which is the first gene to be expressed when the virus infects a plant. The Rep proteins form a multimer, consisting of a number of Rep molecules, which binds to the virus DNA origin of replication. This is a very specific sequence which, because the DNA is single-stranded, forms a hairpin structure. The Rep complex initiates DNA replication at this point, using host proteins to complete the process. Thus, we figured that if the host plant produced many copies of mutated Rep proteins, these would compete with the virus Reps and prevent them from initiating replication.

Tichaona made a variety of mutations in regions of the *rep* gene that are essential for its function, thus producing proteins defective in replication. He tested these in black Mexican sweetcorn (BMS) cells in liquid culture. BMS cells have the advantage of growing in tissue culture almost indefinitely, if treated correctly. As they have many of the attributes of maize plants they can be transiently infected with wild-type MSV and the virus can be challenged with the Rep mutants. Tichaona was able to show that a number of his single and double mutants effectively prevented the virus DNA from replicating in BMS cells. This suggested that if we expressed the Rep mutants in transgenic plants they might make the plants resistant to virus infection.

Wild grasses

Although maize transformation was now a possibility, we knew that it was going to be a slow process and were keen to find a system that would enable us to test Tichaona's constructs rapidly in plants. Ed suggested we try the grass, *Digitaria sanguinalis*, which grew conveniently wild near UCT's tennis courts, as MSV originated in wild grasses. Although leafhoppers prefer to feed on maize, these insects preferentially breed on annual wild grass species (Shepherd et al., 2010). Approximately 70 per cent of the more than 138 grass species on which leafhoppers feed are also MSV hosts, and the density and composition of grass populations in any region almost certainly has a major influence on maize streak disease epidemiology (Varsani et al., 2008).

Sometimes, however, leafhoppers do mate on maize plants, and one of the most fascinating talks I heard on this subject was given by William Page, a scientist from Uganda, at a 1997 international maize streak disease symposium held in Hazyview in Mpumalanga (then the eastern Transvaal). He described an experiment he carried out in maize fields using yellow sticky traps and suction devices. He attached the traps to plants at different heights, and used the suction devices to sample the canopy, especially the whorls. He collected the leafhoppers caught in this way, and sexed them. His interpretation of the results was that the females alight towards the top of plants and the males further down. The females emit some sort of acoustic signal that attracts the males. After a brief mating they both go their separate ways, leaving behind viruses inserted into the plant during the insects' sojourn on the maize. What a legacy of invertebrate courtship!

We accordingly set about developing a transformation system for *D. sanguinalis* which proved to be most successful and took only about four months from bombardment in tissue culture, to adult plants and finally seed (Chen et al., 1998). We were now ready to test Tichaona's genetic constructs, concentrating on the ones that showed promise in his tissue culture experiments.

It was during Tichaona's PhD work that I started to become concerned about a potential competitor in the person of Margaret Boulton at the John Innes Centre in Norwich, UK. She was well known to Ed and had published extensively on MSV. So, believing that it is better to collaborate than to compete, I set off for Norwich with some trepidation. Fortunately, Margaret and I hit it off splendidly and both I and those of my students who spent time in her lab learned a great deal from her, which helped us considerably in our eventual success.

A most important person entered the scene in 1997. Dionne Miles, later Shepherd, did her PhD with Ed and me. She then became a post-doctoral fellow, funded by the Claude Leon Foundation, and is currently employed by UCT, although paid by PANNAR, as a research officer to see the project through to commercialisation. She took over where Tichaona had left off and made more mutations and deletions in the MSV *Rep* gene. Like Tichaona, she first tested them in BMS suspension cells and then used the *D. sanguinalis* transformation system to test them in transgenic plants.

The results were quite variable. Some constructs produced plants showing excellent resistance, even immunity, to MSV but the transgene had deleterious effects on aspects of plant growth and development. Fortunately, some produced healthy, fertile plants with good resistance.

At last we were ready to publish. We submitted a paper entitled 'Maize streak virus-resistant transgenic maize: a first for Africa' to the *Plant Biotechnology Journal*. It came back from the reviewers with requests for corrections. These were accepted and publication was confirmed on 29 June 2007. The date is significant because of what was about to happen.

Publicity

At the beginning of 2007 I had been contacted by Debby Delmer from the Rockefeller Foundation. She had been in charge of the Rockefeller funding for our drought-tolerant maize project (see Chapter 8) for a number of years. She had followed our MSV work with interest and was looking for speakers for the Plant Biology and Botany Joint Congress of the American Society of Plant Biologists to be held in Chicago in July 2007. I recommended she invite Dionne as I figured it was time my post-docs and students got their share of the limelight.

When Dionne presented her paper on July 7 the media immediately jumped upon it and she was interviewed by BBC World and Voice of America. The work was commented upon by *Science* (Figure 7.2) and when the paper was published in November, complete with a front cover picture (Figure 7.1), it created quite a stir in the scientific community, receiving numerous citations. We were relieved that the paper had been accepted before Dionne's presentation in Chicago, as the anti-GMO lobby could well have accused us of 'publishing' in the media before submitting the work to peer review in a scientific journal. The fact that the media came to us, rather than vice versa, would not have entered such a debate!

The work then entered the breeding and glasshouse testing phase at PANNAR, during which the gene was crossed into lines suitable for commercialisation in South Africa and elsewhere on the continent. The resistance has yet to be confirmed by conducting confined field trials.

As with many developers of new technologies, we decided some years ago to have a back-up strategy and that was in the hands of Betty Owor. She was inspired by the work of David Baulcombe (knighted in 2009), famous for his discovery of a phenomenon called gene silencing in plants. This is a process whereby small pieces of RNA can cause genes to be silenced—that is, inactivated. RNA is the molecule that is well known as the intermediate in the DNA → RNA → protein process, immortalised as the Central Dogma by Francis Crick who, together with James Watson, determined the structure of DNA in 1953. These small pieces of RNA, called small interfering (si) RNAs, interfere with the expression of specific genes and can therefore act as antiviral agents. Betty developed just such a system for protecting maize from MSV and this is also being tested both in our labs and at PANNAR. Time will tell which strategy works.

Chapter 8

David vs Goliath

One day I was meeting with some potential funders of my research group and I was explaining to them that, apart from our work on maize resistant to maize streak virus, we were also embarking on tolerance to drought. 'But what if there isn't a drought?' asked one of them, who lives in the UK. People there who knew me well told me they had never before seen me at a loss for words.

Drought is nothing new to sub-Saharan Africa. Newspaper headlines in July 2004 read 'Severe drought depletes SA'. 'The country remains in the grip of one of the worst droughts in recent years, costing the government millions of rands in relief funding' (IOL News, 2004). NASA's earth observatory reported that hot, dry weather from January to March 2007 wilted crops in southern Africa. The severe drought produced near-record temperatures that, combined with a lack of rainfall, caused extensive crop damage. In 2011 the eastern parts of the horn of Africa experienced the worst drought in several decades, resulting in the most severe food security emergency in the world, driven mainly by a combination of food availability and access issues. Two consecutive seasons of significantly below-average rainfall have resulted in failed crop production, depletion of grazing resources and significant livestock mortality (FAO, 2011).

Rockefeller to the rescue

But I need to backtrack to 1996 in order to explain how the funding from the Rockefeller Foundation for our work on drought-tolerant maize began. I received a letter from an unknown PhD student who was on a Fulbright programme at Auburn University in Alabama. His name was Sagadevan Mundree, a former high school teacher from Durban. During the course of his thesis research he had developed an ingenious technique to identify genes that are functionally important in abiotic stress tolerance, using the bacterium *E. coli* to detect genes from any source (Mundree et al., 2000). He now wished to return to South Africa and was looking for a post-doctoral position—could I help? It turned out that he had written to every major university and research institute in South Africa, without success. I immediately wrote back and said he must return as the country needed people like him. I indicated that I would do everything possible to find him a position. And so, in 1997, Saga arrived to take up a post-doctoral fellowship with Jill Farrant in the Botany Department at UCT. The Microbiology Department was able to offer him a job soon after and he joined us in 1999 as a lecturer. Thus began our partnership, which continues to this day.

The first question facing us was where to find interesting genes that might confer tolerance to dehydration, but that other groups weren't working on. This was where Jill Farrant made a strong contribution. She is an expert on a group of plants called 'resurrection plants', which can tolerate high levels of desiccation, losing up to 95 per cent of their water content. They can survive in that dehydrated state for months on end and, when given water, 'resurrect' within 72 hours. For various reasons, Jill suggested we try the South African indigenous species *Xerophyta viscosa* (Figures 8.1 and 8.2).

Figure 8.1   The resurrection plant, *Xerophyta viscosa*,
in its hydrated state

Figure 8.2 *X. viscosa* in its dehydrated state

Where to find these plants? Enter a former post-doctoral fellow, Rachel Saunders, who, with her husband, Rod, ran an indigenous seed nursery. She thought they had seen them growing in the Drakensberg Mountains in KwaZulu-Natal, somewhere near Cathedral Peak. So, with those not-too-specific instructions, Saga applied to the Department of Nature Conservation for a permit. They were extremely helpful and gave us permission to stay at the KwaZulu-Natal Parks Board's Cathedral Peak Nature Reserve's huts, while also supplying us with a guide to help us find the elusive plants. Saga and I set off from UCT with a group of six students, driving a minivan and towing a trailer in the hope of bringing home the loot. We left our base in the foothills of the Drakensberg at first light, climbing up into thick low-lying cloud, not knowing how long it would take us to find our quarry. We emerged into bright sunlight and then, after negotiating a somewhat tricky ledge, were struck with elation as we turned a corner to find an enormous bank covered in glorious *X. viscosa* plants (Figure 8.3).

Each of us filled our backpack with a single plant and lovingly carried it down to camp, where we revelled in the cold mountain stream after our somewhat exhausting eight-hour hike. On a subsequent trip the weather was not quite so kind, but enabled us to see the plants 'resurrecting' before our eyes. There had been a severe drought and

Figure 8.3 *X. viscosa* growing on a rocky slope in the Drakensberg Mountains
near Cathedral Peak

fire in the Drakensberg and the temperatures had been in the high
30s and up to 40°C or more. However, it had rained the night before we
set off and as we reached the ridge in a howling gale, the temperature
was well below freezing. We could indeed see nature at work as the
plants started to change from a stick-like state, becoming healthy
and green. But the going was tough, and when I asked a student
on the way down whether she was pleased she'd come she replied,
'Yeeees, but I'll never do it again!' Fortunately, neither she nor anyone
else ever had to do it again as our technical 'guru', Marion, learned
how to germinate the *Xerophyta* seeds in the glasshouse. However,
I still remember fondly the days when it was a rite of passage for
every graduate student to collect his or her own resurrection plant by
climbing in the Drakensberg.

It was shortly after this initial stage of our research that I met
Gordon Conway, president of the Rockefeller Foundation, at the World
Economic Forum in Davos and our funding started (see Chapter 4).
The Rockefeller Foundation was a marvellous institution to have as a
funder. Once they had decided our project was worthwhile, they gave
us considerable latitude to achieve our objectives. I well remember
going through an annual progress report with the foundations Joe
de Vries, to whom I reported in Nairobi, during a break in an African
Agricultural Technology Foundation board meeting. He commented

that it was probably the best annual report he'd ever read. When I replied that it was only five pages long, he said, 'That's probably why.'

After a few years, Debby Delmer took over from Joe and she wasn't as happy with our progress. She urged us to 'get into maize' as soon as possible. Up until that time we had been content to test our genes in model plants such as tobacco and *Arabidopsis*. However, Debby had burned her hands on projects that got hooked on model plants and never got as far as the crop of interest. Moreover, she was keen that we tried out genes that were unique to *X. viscosa*, not being represented in any of the international gene banks. She suggested I contact Prof Jesse Machuka from Kenyatta University, whom she thought could help us, especially with maize transformation. And that opened up a whole other world.

We had used the 'biolistics' particle bombardment transformation method for introducing genes into maize for the virus resistance project. This was because, at the time we began that project, the method Marc van Montagu and his colleagues used to introduce genes into plants using the bacterium, *Agrobacterium tumefaciens* (see Chapter 7), worked for dicotyledonous plants only—tobaccos and tomatoes—but not cereals such as maize. However, by the early 2000s, *A. tumefaciens* could be used to transform monocots, such as maize, as well as dicots. In many respects it is a more effective method than particle bombardment because it generally delivers a single copy of the gene one wishes to introduce. This has many implications for biosafety and for ease of subsequent breeding into other maize varieties. And this brings me back to Debby's suggestion that I meet Jesse Machuka at Kenyatta University.

Jesse's set-up was inspiring. Apart from a well-equipped laboratory and tissue culture facility, he also had a glasshouse, now a state-of-the-art biosafety-compliant one and, most important of all, his lab sits almost on the equator, making it possible to grow two crops of maize a year. Coming from Cape Town, where we can grow maize only in a temperature- and light intensity-controlled growth chamber, this was little short of miraculous. But best of all, Jesse's lab staff were expert at using *Agrobacterium* to transform maize, and they were as enthusiastic about developing drought-tolerant maize for Africans as I was. And so began our partnership to 'get genes into maize'. The genes would be cloned at UCT and the maize transformed in Kenya. But things like this don't happen overnight.

Slowly our team came together. Dahlia Garwe was the first Zimbabwean to join. She left her husband and two children at home in Harare, and so had a great incentive to work 24/7 in order to return to them as soon as possible. (However, at one point she was forced to return home as her husband, a pilot in the Zimbabwean air force, was required to help in the war in the Democratic Republic of the Congo.) 'Her' gene was Debby's favourite, the one with no similarity to any other in the available databases. There was great excitement when Dahlia found that her gene protected tobacco and *Arabidopsis* (we were still working in model plants at the time) against not only dehydration, but also high temperatures (memories of *Xerophyta* growing at 42°C in the Drakensberg) and salinity. Could we similarly protect maize from a number of different abiotic stresses with Dahlia's gene?

Then came the genes of Alice (she also came from Zimabwe), and of the South Africans, Kershini and Revel. (I have always found it easier to refer to the genes by the student's name, but of course, they all have scientific names as well.) There were others on the way, some of which gave heart-breakingly negative results. For instance, Shaheen, from Mauritius, worked on a very promising gene that coded for an antioxidant protein. As many readers will know, antioxidants are implicated in the ageing process, which can be an extremely stressful condition. But the results sadly showed that this particular antioxidant could not protect plants from environmental stresses.

Thus we honed in on four genes. We had first called Dahlia's gene *XvCor1*, *Xv* for *X. viscosa*, and *Cor* for cold responsive, as we first thought the gene was switched on primarily as a response to low temperatures. When it became clear that it responded to a variety of abiotic stresses we changed the name to *XvSap1*, for stress associated protein. We didn't know how it acted, only that it had six cross-membrane regions (regions of hydrophobicity) and also had some signal-transduction motifs in its DNA, as if parts of it could transmit signals regarding abiotic stresses to the interior of the cell. Alice's was called *XvAld1*, as it codes for the enzyme aldose reductase, which converts glucose to sorbitol, an osmoprotectant. This sugar could act to 'shore up' the interior of the cell in times of osmotic stress due to lack of water. Kershini's was *XvPer2*, being the second antioxidant. This type is called a peroxiredoxin—hence the name. *G6* is another unknown which is induced by cold stress but whose protein levels also increase significantly under dehydration.

However, as we decided on these genes, another strategy emerged. A number of scientists were finding that genes which could potentially protect plants from dehydration could, if switched on continuously, also cause them to grow rather poorly under normal, non-stressed conditions. This is because their expression could result in a physiological load. The situation when a gene is switched on all the time is called constitutive gene expression. The alternative is inducible gene expression, when the gene is switched on only under certain conditions (in this case under conditions of abiotic stress, such as dehydration).

The region of a gene that determines whether it is expressed constitutively or inducibly is called the promoter. This is a region of DNA upstream of the gene which 'instructs' it first to be transcribed into RNA and then translated into the protein that would carry out its function (in our case resistance to dehydration).

Richard Odour was a former USHEPiA PhD student from Jesse's lab. He is now running our research programme at Kenyatta University as a member of their academic staff. His PhD was to develop a promoter that could be used to switch on our four genes when the plant was stressed. Soon after Richard started work, our Rockefeller Foundation grant ground to a halt. All Rockefeller grantees in Debby Delmer's 'biotech' portfolio received a letter from her towards the end of 2006 indicating that no more work on transgenic crops would be funded. We had indeed seen this coming when the new president, Judith Rodin, took over from Gordon Conway. The African Agricultural Technology Foundation had a meeting with its donors and partners at the foundation's offices in New York in 2007, shortly after Judith had taken over. She briefly welcomed us, but it was clear that she was not at all comfortable dealing with an organisation that promoted GM crops. It had been decided that as a gesture from the board, I would present her with a copy of my book *Seeds for the Future*, which I only just managed to do, as she left the room so abruptly after her short interaction with us. Gordon told me later that I was probably not being 'politically correct' when I told her that he had written the foreword.

The South African Maize Trust

As luck would have it, shortly after I had discovered the Rockefeller Foundation as a funding source I also discovered the South African Maize Trust. It was founded in 1998 to promote the maize industry, partly by funding research on crop improvements. I had met a member of a related organisation, Grain SA, at a local conference and when I told him that not only did I not know the Maize Trust existed, I certainly didn't know that they supported research, he encouraged me to apply. Members of the trust visited us a few years after they had started funding our work. They were particularly excited about our use of a South African resurrection plant as a source of genes for drought tolerance and one of them, a farmer from the north of the country, was convinced that he had seen *Xerophyta* plants growing on his farm.

As the project developed, one glaring problem emerged—our glasshouse and growth chambers were totally inadequate for growing maize to the point of setting seed. I had solved this problem with our MSV project by sending the plantlets up to PANNAR Seeds in KwaZulu-Natal where they could set seed, but PANNAR wasn't involved in our drought-tolerance project. We would need to buy a specially controlled environment growth chamber, a Conviron, in which the light intensity could be set to a high enough level for maize to grow properly. And this cost money—a lot of money. UCT made a generous donation, but getting it installed was something of a nightmare. The doors to our building had to be removed and once it was inside and the wooden boards surrounding it removed, we discovered that it had been severely damaged during transit from Canada. Eventually, with the Maize Trust's assistance, everything was in place. It continues to work well, but the space is very small so I was hugely relieved when, in 2010, we eventually transferred all our maize growing to Jesse Machuka's Kenyatta lab in Nairobi.

The other aspect of our research which particularly interested the Maize Trust was our promoter work. From the start of our investigations into drought-tolerant maize, we were aware of the many other groups worldwide working on similar projects. Indeed, it was the large Shinozaki group in the world-renowned RIKEN research institute in Japan who first alerted our attention to the problems of constitutive gene expression in transgenic plants (Kasuga et al., 1999). I've always operated along the lines that it's good to be on friendly terms with the opposition, so on a visit to Japan I spent a day in Prof Shinozaki's lab. Although he had been difficult to talk to at international conferences (the general opinion was that he was not very friendly), on his home turf it was completely different. He couldn't have been more welcoming and, although he didn't share any work that was unpublished, he was perfectly happy to comment on our research. He certainly reinforced the importance of using a stress-inducible promoter.

So Richard Odour began to work on developing the promoter of Dahlia's *XvSap1* gene for use as a signal to switch on a variety of genes and it was this that the Maize Trust decided to patent. Patenting is not for the faint-hearted! I was extremely fortunate to have Revel Iyer working for me. Apart from having a PhD in molecular biology, he also has a law degree and an MBA. Between him and the UCT Patent Office, who together with the Maize Trust footed the bill for the rather costly exercise, the patent was finally filed. Although the Maize Trust will allow farmers in South Africa access, royalty free, to any research they have funded, we all agreed that if commercial farmers in other countries wanted to use it they should pay.

Monsanto, our Goliath

It was around about this time that I became aware of the giant of drought-tolerant maize. I was approached by Monsanto via the African Technology Foundation (AATF), to help publicise their Water Efficient Maize for Africa (WEMA) project. This is funded by the Bill and Melinda Gates Foundation and brings together a number of different organisations. The gene *csp*B has been donated royalty-free by Monsanto and encodes a cold shock protein from the bacterium *Bacillus subtilis* (Castiglioni et al., 2008) which has been shown by them to confer tolerance to dehydration in transgenic maize. The gene has been introduced into African maize varieties by CYMMIT, the international maize and wheat breeding organisation, which, although based in Mexico, also has centres in Africa. These plants are being subjected to confined field trials (CFT) by the National Agricultural Research Services of the partner countries, Kenya, Mozambique, South Africa, Tanzania and Uganda. The project is being managed by the AATF, and Monsanto asked if I would speak on it at an agricultural meeting in Rome. Although the WEMA project is vastly better resourced than ours and far further along the road to farmers' fields, I firmly believe that it is a good idea to have more than one approach to a problem. The WEMA project may deliver everything maize farmers in Africa need for drought alleviation in the years of global warming to come, but it might not. On the other hand, our project might fail completely, but then again, it might not. So I went to Rome and spoke glowingly about the WEMA project, although as I said to my lab, the *csp*B gene in their transgenic maize is, unlike ours, driven by a constitutive promoter. This, then, is the origin of the title of this chapter, David vs Goliath.

I became further involved in Monsanto's WEMA project when, in December 2010, I was asked by the South African Agricultural Research Council (ARC) to speak at a community meeting in Lutzville where the WEMA project was undertaking field trials. Lutzville is a tiny village about four hours' drive from Cape Town up the west coast.

It is a predominantly grape-growing area where little, if any, maize is grown. This, together with the fact that it is usually reliably dry in the summer months, was why it had been chosen as the location for field trials. The ARC wanted me to speak on alternative approaches to drought tolerance, which I thought was pretty broad-minded of this Goliath project. On arrival at the community hall in the town I rapidly discovered two things that the ARC had not communicated to me. Firstly, very few among the gathering of local farmers spoke or understood English (I had never previously given a scientific talk in Afrikaans), and secondly, as growers of mainly organic vegetables (they had found a niche market for themselves), they couldn't care less about maize and even less about GM maize. On the contrary it became increasingly clear that they were extremely angry with the ARC for not caring about their problems, and for foisting on them something they neither needed nor wanted. And, of course, this translated into anti-GM crops sentiment. The over-riding message from this experience was 'do your homework on your target audience' if you don't want the whole exercise to explode in your face.

Two weeks after this meeting the heavens opened up. It didn't just rain on Lutzville; there was a deluge which succeeded in washing away the main (and almost only) road in the town. The WEMA crops survived but their drought tolerance was temporarily irrelevant. Such are the vagaries of working with nature.

Subsequent to this somewhat stormy meeting with community leaders, the possibility of a follow-up stakeholder meeting was discussed. However, the leaders of the group cancelled the meeting and any further engagement with WEMA and later registered an objection to the WEMA trials with the South African regulatory authorities. This reaction was sparked by a public announcement in the daily newspaper that Monsanto intended to carry out another WEMA trial in Lutzville. A public announcement is a requirement by the regulatory authorities for anyone applying for a permit to conduct a GM trial (personal communication, H Mignouna, AATF).

In my opinion this shows the importance of assessing the needs and aspirations of the local farmers, who, in the Lutzville case, had absolutely no interest in maize, and to bear these in mind when holding meetings with them before and during trials. Perhaps the WEMA project was unpopular in Lutzville because the local farming community has extremely deep-seated resentments against what they perceive as the racial agenda of the South African Agricultural Research Council (ARC), the organisation carrying out the trials. In their eyes, under the apartheid regime the old ARC favoured white farmers only. Now the new ARC, under the ANC government, is seen as giving preference to black farmers. The farmers in Lutzville are coloured and feel that once again they are being left out.

However, on a positive note, at a more recent community meeting (November 2011) the ARC invited an expert on vegetable growing and the atmosphere was far more amicable and constructive (personal communication; BioSafety SA).

In the meantime, work continues in Jesse's and my lab on the use of *X. viscosa* genes to confer drought tolerance on African varieties of maize. Richard Odour is back in Jesse Machuka's lab at Kenyatta University, where he is supervising our work. In February 2012 I saw the first developing cobs of 2nd generation GM plants. It is projects such as these that might pave the way to increased acceptance of GM crops in Africa.

Chapter 9

Food for Africa

The InterAcademy Council book entitled *Realizing the Promise and Potential of African Agriculture* (InterAcademy Council, 2004) had stated, as one of its recommendations, that the bridge across the agricultural genetic divide between African countries and those in the developed world needed to be crossed. This divide separates the genetically improved varieties—derived either from conventional breeding or from the various tools of modern biotechnology—available to the developed world from those being used by resource-poor farmers in Africa.

The book concluded that there would need to be substantial investment to respond to the specific needs of African farmers if they were to derive benefit from both conventional breeding and modern biotechnology. Technology needed to be fine-tuned to African needs. We noted the long gestation period biotechnology required before its impact could be realised, and urged for investment sooner rather than later. Some of these calls have been answered and the outcomes will be discussed in this chapter.

Can GM crops help feed the hungry in Africa?

But before I go into these examples, the question must be posed: can GM crops help to feed hungry people in Africa? And if so, will Africans accept GM crops? There are still doubts in certain sectors of African society about the safety of such crops and the foods derived from them. I discussed some of the reasons for this in Chapter 6, specifically identifying the influence that European thinking has on Africans. However, it is essential to know that there has not yet been any documented evidence that approved GMOs have posed any new risks either to human health or to the environment. This is the view not just of the proponents of GM technology, but also of scientific communities and, significantly, even in Europe.

For instance, the Royal Society stated in 2004:

> *We conducted a major review of the evidence about GM plants and human health last year, and we have not seen any evidence since then that changes our original conclusions. If credible evidence does exist that GM foods are more harmful to people than non-GM foods, we should like to know why it has not been made public* (Royal Society, 2004).

Likewise the European Union (EU) Research Directorate wrote in 2001: 'Research on GM plants and derived products so far developed and marketed, following usual risk assessment procedures, has not shown any new risks on human health or the environment.' This statement was based on a study of 81 separate scientific studies conducted over a 15-year period, all financed by the EU rather than private industry, aimed at determining whether GM products were unsafe, insufficiently tested, or under-regulated. It was on the basis of this study that the EU made its statement (Kessler and Economidis, 2001).

Interestingly, in the light of the extreme opposition to GM crops among the French, the French Academy of Medicine and the French Academy of Sciences came out in 2002 with statements confirming the safety of foods derived from such crops. The former announced it had found no evidence of health problems in the countries where GMOs have been widely eaten for several years and the latter stated: 'All the criticisms against GMOs can be set aside based for the most part on strictly scientific criteria.' More recently, in November 2011, France's highest court, the Conseil d'Etat, confirmed that the European Court of Justice's judgement that the 2008 French ban on the cultivation of GM crops was illegal. Both courts overturned the national ban, declaring that the French government presented no scientific evidence of any risk to health or the environment from these crops (http://www.europabio.org).

Commenting on this finding, EuropaBio's director of Green Biotechnology Europe, wrote:

> *These judgments from the highest European court and the highest French court send out a message loud and clear: bans of GM crops cannot be based on political dogma. As both judgments state, no ban on planting GM crops can be declared without valid scientific evidence, something that France and other European countries have not produced* (ibid).

The Union of German Academies of Science and Humanities also expressed their support (Helt, 2004): 'According to present scientific knowledge it is most unlikely that the consumption of the well characterised transgenic DNA from approved GMO food harbours any recognisable health risk.'

More recently a group of European scientists have published an article entitled 'Assessment of the health impact of GM plant diets in long-term and multigenerational animal feeding trials: a literature review' (Snell et al., 2011). The aim of this systematic review was to collect data concerning the effects of diets containing GM maize, potato, soya bean, rice or triticale on animal health. They examined 12 long-term studies, of more than 90 days and up to two years in duration, as well as 12 multigenerational studies from two to five generations. Results from all 24 studies did not suggest any health hazards and, in general, there were no statistically significant differences within parameters observed. What small differences were observed fell within the normal variation range of the considered parameter and thus had no biological or toxicological significance. The studies concluded 'that GM plants are nutritionally equivalent to their non-GM counterparts and can be safely used in food and feed'. Furthermore, they state that a 90-day feeding study performed on rodents, according to the Organisation for Economic Cooperation and Development (OECD) test guidelines, is generally considered sufficient in order to evaluate the health effects of GM feed.

Despite all this overwhelming evidence of the food safety of GM crops, as Robert Paarlberg, the author of *Starved for Science: How Biotechnology is Being Kept Out of Africa* (2008), has written: 'Skeptics who remain fearful sometimes respond that absence of evidence is not the same thing as evidence of absence.' Yet if you look for something for 15 years and fail to find it, that must surely be accepted as *evidence* of absence. It may not be *proof* that risks are absent, but proving something is absent (proving a negative) is known to be logically impossible (Paarlberg, 2012).

It is interesting to compare the acceptance of GM products for human healthcare with the problems relating to GM crops. The cover of *Time* magazine in 1977 showed an evil-looking scientist 'tinkering with life'. A mere four years later Herb Boyer appeared on the cover of the same magazine looking positively cherubic and obviously enjoying the fruits of 'The boom in genetic engineering'. It is clear that when the product is a necessary medicine, people's attitudes are very different from their response to a product that is merely food, which can be taken or left by the consumer. The sad fact is, however, that in some parts of the world, 'mere food' can make the difference between life and death to smallholder farmers who, together with their families, are both producers and consumers.

Environmental effects

If GM crops are safe for humans and animals to eat, what are their effects on the environment? One of the most striking effects has been associated with insecticide and herbicide use. Since 1996, the use of pesticides on crops has reduced by 8.7 per cent, largely due to the planting of GM *Bt* cotton, a crop which, without GM, has traditionally been an intensive user of insecticides. Planting of herbicide-resistant crops has led to a decrease in the overall environmental impact of 16 per cent (Brookes, 2012). This is largely due to a switch to active ingredients with a more environmentally benign profile than the ones generally used in conventional crops. In addition GM crops are contributing to lower levels of greenhouse gas emissions by two principle means. Firstly, there is a reduction in fuel use due to less frequent herbicide and insecticide applications. Secondly, there is a contribution due to the switch to 'no-till' or conservation tilling. Under normal tilling farmers will plough the soil, prior to planting seeds, in order to allow weeds to grow. They then spray with herbicides, but before they can plant they have to allow the toxic chemicals to dissipate, a process which can obviously contribute to the loss of top soil. When planting GM seeds, crops and weeds can grow together and the farmer can spray when he or she sees fit.

Most of these farming practices do not apply to many smallholder African farmers, few of whom can afford insecticides or herbicides. However, there are those who do use pesticides, specifically on cotton, which requires frequent treatment. A study on smallholder farmers in India shows that *Bt* cotton has reduced pesticide applications by 50 per cent, with the largest reductions (70 per cent) occurring in the most toxic types of chemicals. Results confirm that this has notably reduced the incidence of acute pesticide poisoning among cotton growers. These effects have become more pronounced with increasing technology adoption rates. *Bt* cotton now helps to avoid several million cases of pesticide poisoning in India every year, which also entails sizeable health cost savings (Kouser and Qaim, 2011). To quote an African woman farmer from KwaZulu-Natal, farming *Bt* cotton, who addressed the audience at a meeting at the Vatican: 'Look at my hands—they don't look like a farmer's anymore!'

In order for African countries to feel compelled to accept GM crops, such crops must have demonstrated strong agronomic performance, outweighing anything currently available using conventional technologies. It would also help if some of the traits involved improvements to consumers' health. So let's have a look at what's in the pipeline. While doing so we may be able to answer the question I posed above: Can GM crops help to feed hungry people in Africa?

Insect-resistant (*Bt*) cowpea

As discussed in Chapter 4, the African Agricultural Technology Foundation (AATF) is spearheading a project to protect cowpeas from infestation by the borer, *Maruca vitrata*, which can decrease the yield from a potential 2–2.5 to 0.05–0.5 tonnes per hectare (Figure 9.1). The cowpea is one of the most important food legume crops in the semi-arid tropics. This plant is drought tolerant and hence well adapted to the drier regions of the tropics. Being a legume, it fixes atmospheric nitrogen through its root nodules but, unlike many other legumes, its green leaves and pods can also be eaten before crop maturity, which helps to bridge the hunger gap between harvests. It grows well in poor soils with more than 85 per cent sand and with less than 0.2 per cent organic matter and low levels of phosphorus (Singh, 2003). In addition, it is shade-tolerant and can therefore be used in intercropping, a farming method popular in many parts of Africa. Cowpeas can be intercropped with maize, millet, sorghum, sugarcane and cotton and are consumed by approximately 200 million people in Africa. Confined field trials (CFT) have now been carried out in Nigeria and Burkina Faso (Figures 9.2 and 9.3).

The high level of insect pressure introduced into the field trials has so far shown very encouraging results. There is a clear and striking difference between the non-transgenic and the transgenic plants (Figure 9.3). The level of floral, pod and leaf damage is pronounced in the former and no damage has been observed on the latter. Only dead first instar larvae were observed inside the flowers of transgenic events, showing that these plants are resistant to *Maruca* infestation (personal communication: P Addae, AATF).

What are the potential impacts of *Maruca*-resistant cowpea varieties? Baseline studies and ex-ante impact assessments made by the AATF indicate that yields will increase from 0.35 tonnes per hectare (without the use of insecticides) to about 0.5 tonnes per hectare. This represents an anticipated yield increase of 20 to

Figure 9.1   Damage to cowpea caused by the borer *Maruca vitrata*

Figure 9.2   Fenced trial site with a sign for no unauthorised entry at Farako Ba,
Burkina Faso (Source: AATF)

Figure 9.3   Field trials showing (left) transgenic and (right) non-transgenic cowpea plants

40 per cent. Health benefits will accrue from the expected reduction in the use of insecticides and, while difficult to measure, the costs associated with harmful environmental impacts of insecticide use will also decline (AATF Frequently Asked Questions).

Disease-resistant bananas

To most Westerners, bananas are soft and sweet, the so-called 'dessert banana'. However, in Africa the banana cultivars grown are of a firmer, starchier consistency and are called plantains or 'cooking bananas'. They are a major food source and a major income source for smallholder farmers in East Africa. In countries such as Uganda, Burundi and Rwanda, per capita consumption has been estimated at 45 kilograms per year, the highest in the world. I remember during a stay in Uganda reading a newspaper article by a visiting Kenyan journalist in which he wrote: 'If I have to eat another plantain I think I will go mad. Please give me maize!' The difference in food preferences—even among neighbouring African countries—is remarkable.

Cultivated bananas are parthenocarpic, which literally means virgin fruit—the fruit are produced without fertilisation, and are thus seedless and sterile. This has major advantages when it comes to field trials of GM varieties. If there is no possibility of a GM crop cross-pollinating other plants of the same crop, field trials can be conducted in areas where that crop is normally grown. Otherwise, as in the Lutzville field trials of drought-tolerant maize in South Africa (the AATF WEMA project trials), these have to be undertaken in areas where that crop is not grown. Often that puts the trials at a disadvantage, as the climatic and soil conditions at the field site might not be ideal for that crop.

Propagation of bananas involves farmers removing and planting a sucker, a vertical shoot that develops from the base of the plant. If the sucker is removed before it has elongated, the process is even easier, as these suckers can be left out of the ground for up to two weeks.

As discussed in Chapter 4, the AATF is managing a project to protect bananas from the devastating disease called banana bacterial wilt, caused by the bacterium *Xanthomonas campestris*. Prospects of developing varieties with resistance to bacterial wilt through conventional breeding are limited, as no source of germplasm exhibiting resistance against the disease has been identified (personal communication, H Mignouna, AATF). Thus the *Pflp* and *Hrap* genes have been taken from the sweet pepper and introduced into bananas. Confined field trials are being carried out in Uganda using lines that showed enhanced resistance in screen house trials.

Black Sigatoka is a leaf spot disease caused by the fungus, *Mycosphaerella fijiensis*. The disease was first identified in the Sigatoka Valley in Fiji, where an outbreak of this disease reached epidemic proportions from 1912 to 1923 (Marín et al., 2003). It is particularly prevalent in parts of Uganda. When spores of the fungus are deposited on a susceptible banana leaf they germinate within three hours if there is a film of water present or if the humidity is high. This impedes photosynthesis by blackening parts of the leaves, eventually killing the entire leaf. Starved for energy, fruit production falls by 50 per cent or more and the bananas that do grow, ripen prematurely. The fungus is particularly resistant to treatment by antifungal sprays, which, in any case, are usually too expensive for smallholder farmers in East Africa.

As in the above case, the *pflp* and *hrap* defence genes can provide resistance against this pathogen. Under the management of the Ugandan National Agricultural Research Organisation (NARO), transgenic bananas are being assessed for resistance to this fungal disease (personal communication, A Kiggundu, NARO). When the trials began in 2007, they received supportive comment in the journal *Nature* (Dauwers, 2007).

Cassava resistant to cassava mosaic virus (CMV)

Cassava, or manioc, is extensively cultivated as an annual crop in parts of Africa for its edible starchy tuberous root, a major source of carbohydrates, with the crop accounting for up to 30 per cent of the daily calorie intake of Ghanaians. The plant does well in poor soils and with low rainfalls, and, because it is a perennial, it can be harvested as required over time. This wide harvesting window allows it to act as a famine reserve. The importance of cassava to many Africans is epitomised in Ewe (a language spoken in Ghana, Togo and Benin), where the name for the plant is *agbeli*, meaning 'there is life'.

Cassava mosaic virus (CMV) is related to maize streak virus (see Chapter 7) and causes the leaves to wither, limiting the growth of the root. The virus caused a major African famine in the 1920s. Sometime in the late 1980s a mutation occurred in Uganda which made the virus even more harmful, causing the complete loss of leaves. This mutated virus has been spreading at a rate of 80 kilometres per year. I realised the danger of this disease when I first met Betty Owor at the Ugandan branch of the International Institute for Tropical Agriculture. She was to become an important player in my maize streak virus resistance team at UCT, bringing her expertise on the CMV to bear on our problem. She pointed to a map on the laboratory wall which showed the encroachment of CMV into Nigeria via the Democratic Republic of Congo and neighbouring countries. Nigeria, as the world's largest producer of cassava, was starting to become seriously concerned.

CMV-resistant cassava is being tackled by Claude Fauquet and his team at the Donald Danforth Plant Research Institute in St Louis, Missouri. They are using RNA interference (RNAi) technology via the replication associated *AC1* gene from the East African strain of CMV isolated in Uganda. This technology can be described as a natural defence mechanism of plants and other organisms, and consists of 'teaching' the plant to recognise virus genetic sequences in advance, so that the plant is ready to act when the real virus attacks. Indeed, it can be thought of as a type of vaccination.

The transgenic lines have been tested extensively under glasshouse conditions in the US. The National Agricultural Research

Organisation of Uganda is now undertaking confined field trials in Namulonge, a known hotspot for CMV (personal communication; A Kiggundu, NARO).

Drought-tolerant maize

The WEMA (Water Efficient Maize for Africa) project has been extensively covered in Chapter 8, but recently three additional potential open quarantine sites for carrying out confined field trials under random stress in drought-prone areas have been identified in the Limpopo Province in South Africa. These are in a maize-growing area of the country where local farmers will be keenly interested in the outcome of the trials. This is in contrast to the trials carried out in Lutzville, a grape-growing area in the Western Cape. Possibly the new trials in Limpopo, where maize is an extremely important crop, will have a better chance of public acceptance.

WEMA trials are being carried out in most of the partner countries. In addition to those in South Africa, trials are underway in Kenya and Uganda, and sites for future testing have been identified in Mozambique and Tanzania. By 2010 the partners had earmarked 12 hybrid drought-tolerant maize varieties developed through the project for CFTs. But, as Dr Sylvester Oikeh, the AATF project manager wrote in their annual report: 'Our application for permits to conduct CFTs in Kenya, Uganda and Tanzania progressed much slower and in a more complex way than we would have wished. However, we worked closely with the regulatory authorities in the different countries and were vigilant in their requirements.'

The process was aided by mock trials conducted by the WEMA partners in Kenya and Tanzania under the supervision of biosafety inspectors. The officials then monitored the harvested crop at two-week intervals until they were satisfied that all post-harvest requirements had been satisfied. As a result, CFTs have been undertaken in Uganda and Kenya, as well as in South Africa. Tanzania and Mozambique have yet to approve trials in their countries and the jury will need some years of data before they can come to a verdict on the effectiveness of the *cspB* gene.

Improved rice varieties

One doesn't readily think of rice and Africa in the same breath, but it is an important staple food and a commodity of strategic significance across much of the continent, but particularly in humid West Africa. The demand for rice is growing faster in this region than anywhere else in the world (http://www.scienceinafrica.co.za/nerica.htm: accessed 29 October 2011). However, the yield is low, due in part to abiotic factors such as nitrogen deficiency and drought, while high salinity is also a problem in many rice-growing areas.

To overcome these problems the AATF is coordinating a multiple-partner collaboration involving Arcadia Biosciences in California, the Public Intellectual Property Resource for Agriculture (also based in California) the International Centre for Tropical Agriculture in Colombia and the African National Agricultural Research Systems in the countries involved. This work is still in the laboratory stage where transgenic plants are being tested for homozygosity of the genes ready for shipment to countries such as Ghana.

Striga-resistant maize

Striga is a major parasitic weed that infests about 20 million hectares of arable land in sub-Saharan Africa. As discussed in Chapter 4, the AATF has facilitated the delivery of StrigAway® seeds, coated with the herbicide Imazapyr, to farmers in Kenya and Tanzania, where commercial seed production began in 2006 and 2010 respectively. Variety testing is ongoing in Uganda and Nigeria. One of the reasons for the early success of this product is that it is not a GM variety, hence none of those attendant regulatory hurdles had to be met. However, a number of useful lessons were learned from this exercise:

(1) The capacity of seed companies to produce sufficient quantities of certified seed was limited due to inadequate land, irrigation and seed coating facilities.
(2) Many agro-dealers near the *Striga*-infested farmlands have little capital, and hence reduced seed-stocking capacity.
(3) Even though there are no GM regulations to deal with, there are elaborate herbicide registration requirements and these may delay the sale of StrigAway® seed.

To overcome these problems the following steps need to be taken:

(1) A credit access system for seed companies, agro-dealers and farmers is required.
(2) Strategic outreach and awareness programmes are needed to disseminate information on new varieties.

(3) Linkages are required involving both private and public sector institutions—for example, seed banks, seed companies and NGOs (personal communication, H Mignouna, AATF).

What are the anticipated impacts of *Striga*-resistant maize which has been named 'UaKayongo', Swahili for 'kill the Striga weed'? The AATF estimates that when fully adopted in Kenya, where about 210 000 hectares of land are infested with the weed, it will lead to an extra 62 000 million tonnes of maize with a value of about US$11.2 million at 2011 prices. Data from on-farm trials indicate that average yields due to the improved seeds have increased from 500 kilograms per hectare to 3000 kilograms per hectare. The expected net benefit-to-cost ratio for use of StrigAway® seeds for an average farmer is around 45:1, a return of 45 per cent. Moreover, because using the technology reduces the weed seed bank over time, abandoned farmland can be recovered and once again cultivated. The results being achieved in Kenya could be replicated in other countries where *Striga* has similar negative impacts on crop productivity (AATF Frequently Asked Questions).

Vitamin-enhanced crops

Vitamin A deficiency is common in developing countries. One of its earliest manifestations is night blindness, which is often found in malnourished pregnant women and children. Many of these children die within a year of becoming blind. In an effort to help overcome this problem in Asian countries where rice is the staple food, Ingo Potrykus and Peter Beyer genetically engineered a variety of rice that produced beta-carotene, a precursor of vitamin A. The first version produced only low levels of the micronutrient, but a subsequent version contains sufficient amounts to provide the entire dietary requirement of the nutrient to people who eat about 75 grams of this Golden Rice per day (Paine et al., 2005). It is called Golden Rice because beta-carotene has a yellow colour.

Many people who oppose the use of GM crops cite patents as one of the stumbling blocks in their development and deployment to poor farmers. Ingo Potrykus said that if his team had not been able to 'piggy back' on the research covered by these patents they would never have been able to develop Golden Rice, certainly not in the time that it took them.

This was fine for the research; when it came to deployment that was another matter. The team had received funding from the European Commission's 'Carotene Plus' research programme and they were required by law to give the rights to their discovery to the corporate sponsors, Syngenta. Free licences were required from all these patent holders so that Syngenta and the humanitarian partners in the project could use Golden Rice in breeding programmes for release (Potrykus, 2001). Fortunately, these were quickly granted due to the positive publicity Golden Rice received, as it was said to be the first GM crop that had benefits for the consumer, not just the farmer. Monsanto was one of the first to grant free licences.

So why is Golden Rice not available for vitamin A-starved children in Asia? Anti-GM crop movements, in particular Greenpeace, have effectively blocked it for all these years. Speaking at a Manitoba Special Crops Symposium in February 2012, Dr Patrick Moore, a co-founder of Greenpeace, explained why he is particularly concerned about Greenpeace's success in blocking its introduction. 'GM rice varieties are able to eliminate micronutrient deficiency in rice-eating countries, which afflicts hundreds of millions of people, and actually causes between a quarter and half a million children to go blind and die young each year because of vitamin A deficiency because there is no beta-carotene in rice. We can put beta-carotene in rice through genetic modification, but Greenpeace has blocked this.' (http:// www.biotech-now.org/food-and-agriculture/2012/02/greenpeace-founder-biotech-opposition-is-crime-against-humanity?utm_source=Enewsletter&utm_medium=Email&utm_campaign=BIOtechNOW accessed 28 March 2012).

Sorghum

How to deal with vitamin A deficiency in Africa? Golden maize would not be an option because, throughout Africa, yellow maize is used as fodder for livestock, while people eat white maize. Why not use sorghum as a source of this micronutrient? In terms of tonnage, sorghum is Africa's second most important crop. The continent produces about 20 million tonnes per annum—about one third of the world crop.

However, these figures do not do justice to the importance of sorghum in Africa. It is the only viable food grain for many of the world's most food insecure people, and is uniquely adapted to Africa's climate, being both drought resistant and able to withstand periods of water-logging. Sorghum originated in Africa and Africans know how to plant, cook and eat it. It is processed into a wide variety of attractive and nutritious traditional foods, such as semi-leavened bread, couscous, dumplings and fermented and non-fermented porridges. New sorghum products, such as instant soft porridge and malt extracts, are proving to be great successes. In the competitive environment of multinational enterprises, sorghum has proven to be the best alternative to barley for lager beer brewing (http:// biosorghum.org/articles.php?id=80).

Florence Wambugu, Head of Africa Harvest, set up the African Biofortified Sorghum project consortium to work on the problem of vitamin enrichment. The beauty of this consortium is that it leverages the best private and public partnerships to deliver the technology as follows:

- The national agricultural research institutes in the partner countries in East and southern Africa bring their expertise in field trials and breeding.
- Pioneer donated the technology, estimated to be worth US$ 4.8 million. They invested in African capacity building, ensuring Africa's contribution would not be token, but strengthened for future sustainability. Over 70 scientists have been involved in the project.
- The technology and research organisations in the partner countries became the African technology recipients to enhance and customise the technology in intellectual property for use in Africa.
- The universities involved added infrastructure and human resources for the analytical work involved.
- Institutions such as the AATF, Africa Harvest and WECARD (West and Central African Council for Agricultural Research and Development) help influence national policies across country borders through advocacy for stakeholder awareness and technology acceptance (http://biosorghum.org/articles. php?id=74).

As discussed in Chapter 2, the first glasshouse trials of the transgenic plants were to have been performed in South Africa but the regulatory authorities refused permission. As a result the trials and further work are being carried out in Kenya.

In the hands of the politicians

With all these exciting developments in the pipeline, together with the already established GM crops widely grown in South Africa (insect- and herbicide-resistant maize, insect-resistant cotton and herbicide-resistant soya beans), I think the answer to my earlier question 'Can GM crops help to feed hungry people in Africa?' is a resounding 'Yes'. However, there still remains the question: Will GM crops help to feed hungry people in Africa? To a great extent that depends on the politicians.

As we have seen throughout this book, and as was so clearly seen in the case of Kenya regarding the Biosafety Act and Regulations (see Chapter 6), politicians can make or break this technology. I was recently contacted by my former PhD student, Dahlia Garwe (see Chapter 8), now acting general manager of the Zimbabwe Tobacco Research Board. She had heard from the Zimbabwean Secretary for the Ministry of Science and Technology Development, Prof Gudyanga, who had returned from an African Development Bank meeting in Nairobi in February 2012. He was fired up by the possibilities that GM crops could hold for his country, despite the Minister of Agriculture having given it a resounding 'NO'. She told me that she had been asked to produce a paper for cabinet on the subject and was asking for 'ammunition'. Time will tell the outcome of this.

Robert Paarlberg also discusses the role of politicians in his book *Starved for Science: How Biotechnology is Being Kept Out of Africa* (Paarlberg, 2008). He concludes as follows: 'So in the end it is not the citizens of Africa who are rejecting agricultural biotechnology. The technology is being kept out of Africa by a careless and distracted political leadership class that pays closer attention to urban interests and to inducements from outsiders ... than to the needs of their own rural poor.'

If this continues to be the case it could indeed be a tragedy for the rural poor. Food riots have been hitting African countries such as Cameroon, Burkina Faso and Mozambique since 2008. Later they spread to Haiti, Bolivia and even Israel and Iceland. The world is simply going to have to wake up and realise that more food will have to be produced on the same amount of land at cheaper costs. And one of the ways to do this is to use technologies that enable farmers to grow better crops. One day, perhaps, even Europeans will wake up and realise that their perception that GM crops put money only into the pockets of seed companies and farmers and not into the mouths of consumers is not necessarily true.

The views of Africans themselves

But what do the African people themselves think? What are the sociological factors influencing the adoption of GM crops in sub-Saharan Africa? A group of researchers recently asked this question of 91 people in five African countries. They found that there were four recurring factors: communication, culture and religion, capacity building and commercialisation (Ezezika et al., 2012).

Poor communication has led to a limited understanding of GM crops by the public. One stakeholder is quoted as saying: 'My understanding is that a number of people, including politicians and some decision-makers, do not know really what GM is.' There is a feeling that communication both from the media and from researchers, especially to grassroots communities, tends to be elitist—'a little above the common man'. A need for 'barefoot extension officers' was suggested. Attention was drawn to the fact that anti-GM crops interest groups have the capacity for widespread dissemination of information at the grassroots level and can spread misinformation, creating extensive public concern and distrust for agbiotech initiatives.

Another common theme is the importance of cultural and religious issues. The different roles of women and men in agriculture were highlighted. Most farmers are women, yet women have limited decision-making roles. Women are the primary agriculturalists where men are involved in the secondary work of buying and selling agricultural products. Men decide on new technologies but it is the women who must implement them. Stakeholders indicated that for agbiotech to be successful, changes must be made to the current system to involve women in leading the decision-making process to ensure that GM products reach women farmers and consumers.

Linked to the theme of cultural issues was the concern that modern agbiotech practices and business models could adversely affect traditional seed systems, including seed selection and breeding, seed sharing and lead to the loss of indigenous varieties. However, other stakeholders hoped that the new agbiotech approaches would be adaptable to traditional seed systems and provide safeguards to traditional seeds that have been cultivated over the years. On the religious side, participants described the perceptions held by many Africans that agbiotech is unnatural, as interfering with nature. Genetic engineering may be regarded by some as taking on the role of God.

Regulatory officials in Kenya, Uganda, Mozambique and Tanzania, in particular, identified inadequate training and expertise as a major constraint. They saw the urgent need for capacity building at the graduate and postgraduate levels of people with expertise in biotechnological applications. This was linked to the need for more local product development. Some participants felt that products were being imposed upon them by Americans and Europeans. In contrast, a product developed within the country is considered more acceptable. Indeed some considered agbiotech as being an 'effort by the Western world to come and take advantage of poor Africans'. Distrust of the private sector, particularly multinational companies, is a major factor in the resistance to agbiotech adoption.

Finally, on the issue of commercialisation of GM crops, it was agreed that there was a need to see some sort of benefit, in yield, health outcomes or other tangible advantages, as a key factor in adoption. One stakeholder is quoted as saying: 'If the (agricultural) technology is more efficient, it will be adopted. Farmers are not sentimental.' This harks back to the point made by Paarlberg that in order for a GM crop to be accepted its advantages must outweigh anything currently available using conventional technologies. In addition, the products of GM crops must be culturally appropriate in terms of appearance, taste, texture, processing qualities and storability.

While considering the advantages of GM crops we must also take into account potential disadvantages to the farmer. One that springs to mind is that the use of *Bt* crops could result in the development of insects that are resistant to this toxin. This is by no means a new concern. Indeed, it was expressed by both opponents and proponents of insect-resistant crops from their earliest deployment. Farmers are well aware that insects can develop resistance to chemical pesticides—so why not to *Bt* crops? Constant exposure to any toxin creates evolutionary pressure for pests to become resistant. One method of reducing resistance is to create non-*Bt* crop refuges to allow some insects to survive, even though they are susceptible, and thus maintain a susceptible population, so that any resistance genes that arise will become greatly diluted. Thus farmers buying *Bt* maize seeds are informed of the importance of planting fields of non-transgenic plants, refuges adjacent to fields of the *Bt* variety.

Is this working? On the whole, yes, but pockets of resistance have been noted, one being in South Africa. Recently AfricaBio, together with Crop Life International, held a workshop to discuss this problem. Delegates came from six African countries: Burkina Faso, Ghana, Kenya, Nigeria, Uganda and South Africa. Clearly this is a problem they wish to avert and they concluded that an effective and functional insect-resistance management system for Africa is essential if the continent is to derive maximum benefit from this technology.

In 2010 I was surprised to be invited to be a keynote speaker at a 'summit' on climate change held by the iLembe district municipality in KwaZulu-Natal. I was surprised because firstly, they had chosen to hear about the role that GM crops could play in climate change, but secondly, they had specifically asked me to talk on our work on the development of maize resistant to maize streak virus. This is a sugarcane-growing region.

I stood up to speak with some trepidation, not knowing what kind of reception I would receive from this very local audience, but I needn't have been concerned. The response was overwhelmingly positive, with one of the loudest refrains being: 'Can this technology also be applied to sugarcane? And if so let's make it happen!' I left feeling really buoyant because I realised that ordinary citizens do still have a voice when it comes to their own interests. And if they feel strongly enough about this new technology it does indeed have a chance of happening in Africa.

Glossary

Anaerobe: organism capable of growing in the absence of oxygen

ATP: adenosine-5'-triphosphate, the molecule that provides energy to an organism

Bacteriophages: viruses which grow in bacteria

Bacteriophage vectors: bacterial viruses able to introduce foreign genes into bacteria

Biolistics: the process whereby genes are 'shot' into plant cells using a device powered by helium gas; the word 'biolistics' is derived from the words 'biological' and 'ballistics', as early versions of the apparatus looked like guns and were powered by gunpowder; the device is still called a 'gene gun'

Callus: undifferentiated plant cells which can be transformed with DNA and regenerate into whole plants

Clone: an identical copy of something [from the Greek word for 'twig' or 'slip']

Collagenolytic: able to break down collagen, the main protein of animal hides

Constitutive gene expression: the situation when a gene is switched on to make RNA and protein continuously

Cyclotron: an apparatus in which charged atomic and subatomic particles are accelerated by an alternating electric field while following an outward spiral or circular path in a magnetic field; this can be used to introduce mutations

Dicot/dicotyledonous plant: a flowering plant with an embryo that bears two cotyledons (seed leaves), typically with broad, stalked leaves with netlike veins

Eukaryotes: organisms containing 'true nuclei', ranging from yeasts to humans, as opposed to prokaryotes, organisms containing 'pre-nuclei', such as bacteria

Event: a single genetic transformation result whereby a specific gene, or set of genes, is introduced into a plant variety

Gene bank: a 'bank' of hybrid plasmids carrying every gene from a given organism

Genetic engineering: the science of introducing foreign genes into an organism, be it a bacterium, a plant or an animal

Hybrid plasmid: a plasmid carrying one or more additional genes and able to be transferred into a bacterial strain, in this case *E. coli* lacking the *Pfk* gene

Inducible gene expression: the situation when a gene is only switched on under certain conditions

***In vitro* transcription/translation extract:** an extract of *E. coli* cells that is capable of supporting the transcription and subsequent translation of one or more genes

Liquid scintillation counter: a machine that measures the radioactivity of all the samples placed in it and provides a printout of the results

Monocot/monocotyledonous plant: a flowering plant with an embryo that bears a single cotyledon (seed leaf), typically with elongated stalkless leaves with parallel veins

Osmoprotectant: a compound that protects an organism from osmotic shock

Phosphorylation: introduction of a phosphoric group into a molecule, a process which traps energy for use by the organism

Plant vectors: agents, such as bacterial vectors, that can introduce foreign genes into plants

Plasmid: a small circle of DNA that is capable of replicating independently of the chromosome in a bacterium and can be used as a vector to introduce genes into other bacteria

Recombinant DNA molecule: any fragments of DNA from the same or different species that have been purposefully recombined to generate a new DNA molecule

Restriction enzyme: an enzyme that cuts DNA at specific sequences, called restriction sites, very often producing single-stranded (sticky) ends

Stem cells: undifferentiated cells of a multicellular organism that are capable of giving rise to indefinitely more cells of the same type

Totipotent: the ability of a single cell to form a complete organism; a single plant cell can develop into an adult plant because its cells are totipotent

Transcription: the conversion of DNA into RNA

Transformation: the introduction of a foreign gene into a plant

Transgenic plant: any plant carrying foreign gene(s) made by genetic engineering

Translation: the conversion of RNA into protein

Bibliography

AATF Annual Report (2010). ISSN 1817–5813.

AfricaBio *Agricultural Biotechnology: Facts for Decision-Makers* and *Biotechnology: Biosafety, Food Safety and Food Aid* (selected articles):

The Impact of Biotechnology on Africa in the 21st Century (June 2001).

China surges ahead of India in Biotech race (February 2002).

Zambia launches its first biotech outreach society (July 2003).

SA GMO maize crops set to grow (April 2004).

International pressure group Greenpeace warns Philippine authorities that biotechnology 'can lead to millions of dead bodies, sick children, cancer clusters and deformities' (April 2004).

Tanzania jumps on GM bandwagon—Agricultural Ministry says they cannot afford to be left behind (March 2005).

Golden rice provides increased Vitamin A (March 2005).

Kenyan Minister asks journalists to highlight biotech benefits (June 2006).

UK farmers optimistic about GM crops (February 2008).

Bt toxin resistance: an evolutionary action (March 2008).

Consumer Protection Regulation effective October 2011 (October 2011).

Partners host successful IRM workshop (December 2011).

African Biosafety Network of Expertise (2012). Follow-up on controversial GM maize study by Seralini et al. 'European Food Safety Authority confirms their conclusion that the study is not valid for risk assessment'. ABNE, Burkina Faso.

Agricultural Business Chamber (2011). www.agbiz.co.za

Aldridge S (1996). *The Thread of Life: The Story of Genes and Genetic Engineering*. UK: Cambridge University Press.

A National Biotechnology Strategy for South Africa (2001). Department of Arts, Culture, Science and Technology, p 4.

Ammann K (2007). 'Reconciling traditional knowledge with modern agriculture: a guide for building bridges'. In A Krattiger, RT Mahoney, L Nelsen, JA Thomson, AB Bennett, K Satyanarayana, GD Graff, C Fernandez and SP Kowalski (eds) *Intellectual Property Management in Health and Agricultural Innovation: A Handbook of Best Practices*. Oxford, UK: MIHR and Davis, California, USA: PIPRA pp 1539–1559.

Avise JC (2004). *The Hope, Hype, and Reality of Genetic Engineering: Remarkable Stories from Agriculture, Industry, Medicine, and the Environment*. UK: Oxford University Press.

Beever DE and Kemp CF (2000). 'Safety issues associated with the DNA in animal feed derived from genetically modified crops. A review of scientific and regulatory procedures'. *Nutrition Abstracts and Review. Series A, Human and Experimental*. 70: 197–204.

Berg P, Baltimore D, Boyer HW, Cohen SN, Davis RW, Hogness DS, Nathans D, Roblin R, Watson JD, Weissman S and Zinder ND (1974). Potential biohazards of recombinant DNA molecules. *Science* 185: 303. http://beck2.med.harvard.edu/week13/The%20Science%20Letter.pdf

Biosorghum
    http://biosorghum.org/articles.php?id=80 (accessed 13 March 2013).
    http://biosorghum.org/articles.php?id=74 (accessed 13 March 2013).

Biotech Now
    http://www.biotech-now.org/food-and-agriculture/2012/02/greenpeace-founder-biotech-opposition-is-crime-against-humanity?utm_source=Enewsletter&utm_medium=Email&utm_campaign=BIOtechNOW (accessed 28 March 2012).

Birner R, Abel Kone S, Linacre N and Resnick D (2007). 'Biofortified foods and crops in West Africa: Mali and Burkina Faso'. *AbBioForum* 10(3): 192–200. http://www.agbioforum.org (accessed 13 March 2013).

Boadi, RY and Bokanga M (2007). 'The African Agricultural Technology Foundation approach to IP management'. In A Krattiger, RT Mahoney, L Nelsen, JA Thomson, AB Bennett, K Satyanarayana, GD Graff, C Fernandez and SP Kowalski (eds). *Intellectual Property Management in Health and Agricultural Innovation: A Handbook of Best Practices*. Oxford, UK: MIHR and Davis, California, USA: PIPRA pp 1765–1774.

Brookes, G and Barfoot P (2011). *GM crops: global socio-economic and environmental impacts 1996–2009*. PG Economics, UK. http://www.pgeconomics.co.uk/publications.php (accessed 13 March 2013).

Branch, M and Branch G (1981). *The Living Shores of Southern Africa.* Cape Town: Struik.

Brookes, G (2012). Genetically engineered crops: environmental impacts 1996–2009. ISB News Report January 2012.

Bull, AT, Holt G and Lilly MD (1982). *Biotechnology: international trends and perspectives.* Paris: Organisation for Economic Co-operation and development. http://www.oecd.org/dataoecd/34/9/2097562.pdf

Business Day 15 October (2002).

Carson, R (1962). *Silent Spring.* New York: Houghton Mifflin.

Castiglioni P, Warner D, Bensen RJ, Anstrom DC, Harrison J, Stoeker M, Abad M, Kumar G, Salvador S, D'Ordine R, Navarro S, Back S, Fernandes M, Targolli J, Dasgupta S, Bonin C, Luethy MH and Heard JE (2008). 'Bacterial RNA chaperones confer abiotic stress tolerance in plants and improved grain yield in maize under water-limited conditions'. *Plant Physiology* 147: 446–455.

Chen CH, Lin HJ, Ger MJ, Chow D and Feng TY (2000). 'cDNA cloning and characterization of a plant protein that may be associated with the hairpinPSS-mediated hypersensitive response'. *Plant Molecular Biology* 43: 429–438.

Chen W, Lennox SJ, Palmer KE and Thomson JA (1998). 'Transformation of *Digitaria sanguinalis:* a model system for testing maize streak virus resistance in Poaceac'. *Euphytica* 104: 25–31.

Conway G (1997). *The Doubly Green Revolution: Food for All in the 21st Century.* UK: Penguin.

Conway G (2003). 'Sowing the seeds of modified crops'. *Nature* 421: 478–479.

Conway G (2012). *One Billion Hungry: Can We Feed the World?* Ithaca and London: Cornell University Press.

Cotton SA. Cotton market report. http://www.cottonsa.org.za (accessed 13 March 2013).

Dauwers A (2007). 'Uganda hosts banana trial'. *Nature* 447: 1042.

Diamond J (1998). *Guns, Germs and Steel.* New York: NW Norton.

Douches DS, Li W, Zarka K, Coombs J, Pett W, Grafius E and El-Nasr T (2002). 'Development of *Bt-cry5* insect-resistant potato lines Spunta-G2 and Spunta-G3'. *Horticultural Science* 37: 1103–1107.

Earth Observatory, March 1 (2007). http://earthobservatory.nasa.gov/NaturalHazards/view.php?id=18226 (accessed 26 March 2012).

EuropaBio   http://www.europabio.org/agricultural/press/highest-courts-france-and-eu-confirm-france-s-ban-gm-crops-illegal (accessed 13 March 2013).

European Union (EU) Research Directorate (2001). GMOs: are there any risks? Brussels: EU Commission, press briefing, 9 October.

Evans JH (2002). *Playing God?: Human Genetic Engineering and the Rationalization of Public Bioethical Debate.* USA: University of Chicago Press.

Ezezika OC, Daar AS, Barber K, Mabeya J, Thomas F, Deadman J, Wang D and Singer PA (2012). Factors influencing agbiotech adoption and development in sub-Saharan Africa. *Nature Biotechnology* 30: 38–40.

FAO Emergency Ministerial-level meeting, 25 July 2011.

Farm and Ranch Guide
http://www.farmandranchguide.com/news/regional/carlson-representing-agriculture-at-the-the-world-economic-forum/article_56324758-50e0-b7cd-001871e3cebc.html (accessed on 10 February 2012).

French Academy of Medicine (2002). 'OGM et santé. Recommendations' (Alain Rerat). Communiqué adopted on 10 December.

French Academy of Sciences (2002). Genetically modified plants. Institut de France, Academie des sciences, Report on Science and Technology, 13 December 2002.

Fuller, C (1901). 'Mealie variegation'. In 1st Report of the Government Entomologist, Natal, 1899–1900.

Giddings LV, Potrykus I, Ammann K and Fedoroff NV (2012). 'Confronting the Gordian knot'. *Nature Biotechnology* 30: 208–209.

Gordon-Kamm WJ, Spencer TM, Mangano ML, Adams TR, Daines RJ, Start WG, O'Brien JV, Chambers SA, Adams Jr WR, Willetts NG, Rice TB, Mackey CJ, Krueger RW, Kausch AP and Lemaux PG (1990). 'Transformation of maize cells and regeneration of fertile transgenic plants'. *The Plant Cell* 2: 603–618.

Gouse M (2012). 'Farm-level and socio-economic impacts of a genetically modified subsistence crop: the case of smallholder farmers in KwaZulu-Natal, South Africa'. PhD thesis, University of Pretoria.

Gouse M, Kirsten J and Jenkins L (2003). '*Bt* cotton in South Africa: adoption and the impact on farm incomes amongst small-scale and large-scale farmers'. *Agrekon* 42:15–28.

Greenpeace
http://www.biotech-now.org/food-and-agriculture/2012/02/greenpeace-founder-biotech-opposition-is-crime-against-

humanity?utm_source=Enewsletter&utm_medium=Email&utm_campaign=BIOtechNOW (accessed 28 March 2012).

Hardie JA (1977). 'Biology's "atomic bomb"'. *NRTA Journal*, January–Febuary 1977.

Heidhues F, Atsain A, Nyangito H, Padilla M, Ghersi G and Le Vallée J-C (2004). *Development Strategies and Food and Nutrition Security in Africa: An Assessment*. Washington, DC: IFPRI.

Helt, HW (2004). 'Are there hazards for the consumer when eating food from genetically modified plants?' Union of the German Academies of Science and Humanities, Commission on Green Biotechnology. Gottingen: Universitat Gottingen.

Ho, M-W (2000). *Genetic Engineering—Dream or Nightmare: Turning the Tide on the Brave New World of Bad Science*. New York, NY: Continuum. http://www.ipetitions.com/petition/changeeugmlegislation (accessed 25 October 2011).

Huang SN, Chen CH, Lin HJ, Ger MJ, Chen ZI, and Feng TY (2004). 'Plant ferredoxin-like protein AP1 enhances Erwinia-induced hypersensitive response of tobacco'. *Physiological and Molecular Plant Pathology* 64: 103–110.

Hubbell S (2002). *Shrinking the Cat: Genetic Engineering Before We Knew About Genes*. USA: Houghton Mifflin.

IFOAM (2005). *Definition of Organic Agriculture*. Bonn: IFOAM (International Federation of Organic Farming Movements) proposals p 12.

IFPRI (International Food Policy Research Institute) (2002). *Reaching Sustainable Food Security for All by 2020. Getting the Priorities and Responsibilities Right*. Washington, DC: IFPRI.

IFPRI (2012). A 'State of Affairs' assessment of agricultural biotechnology for Africa. (still in draft form).

InterAcademy Council (2004). *Realizing the Promise and Potential of African Agriculture*. Amsterdam: Inter Academy Council.

IOL News, 27 July 5 (2004). http://www.iol.co.za/news/south-africa/severe-drought-depletes-sa-1.216364 (accessed 26 March 2012).

James C (2011). 'Global status of commercialized biotech/GM crops: 2011'. *ISAAA Brief* no. 43. Ithaca, NY: ISAAA.

James C (2012). 'Global status of commercialized biotech/GM crops: 2012'. *ISAAA Brief* no. 44. Ithaca, NY: ISAAA.

Jaroff L, Golden F and Jucius A (1977). 'Tinkering with life'. *Time* April 18: 50–55.

Karembu M, Otunge D and Wafula D (2010). *Developing a Biosafety Law: Lessons from the Kenyan Experience*. Nairobi, Kenya: ISAAA *Afri*Center.

Kasuga M, Liu Q, Miura S, Yamaguchi-Shinozaki K and Shinozaki K (1999). 'Improving plant drought, salt, and freezing tolerance by gene transfer of a single stress-inducible transcription factor'. *Nature* 17: 287–291.

Kenward M (1989). 'Let's get back to basics, says Thatcher'. *New Scientist* 124: 13.

Kessler C and I Economidis (eds) (2001). *EC-sponsored Research on Safety of Genetically Modified Organisms: A Review of Results.* Luxembourg: Office for Official Publications of the European Communities.

Kouser, S and Qaim M (2011). 'Impact of *Bt* cotton on pesticide poisoning in smallholder agriculture'. *Ecological Economics* 70: 2105–2113.

Krattiger A, Mahoney RT, Nelsen L, Thomson JA, Bennett AB, Satyanarayana K, Graff GD, Fernandez C and Kowalski SP (eds) (2007). *Intellectual Property Management in Health and Agricultural Innovation: A Handbook of Best Practices.* Oxford, UK: MIHR and Davis, California: PIPRA pp 1539–1559.

Kuyek D (2002). 'Intellectual property rights in African agriculture: implications for small farmers'. Genetic Resources International (GRAIN). 2 August.

Lambrecht B (2002). *Dinner at the New Gene Café: How Genetic Engineering is Changing What We Eat, How We Live, and the Global Politics of Food.* USA: St Martin's Press.

Lutjeharms J and Thomson JA (1993). 'Commercializing the CSIR and the death of science'. *South African Journal of Science* 89: 8–14.

Marín DH, Romero RA, Guzmán M and Sutton TB (2003). 'Black sigatoka: an increasing threat to banana cultivation'. *Plant Disease* 87: 208–222.

McHughen A and Smyth S (2008). 'US regulatory system for genetically modified organisms, rDNA or transgenic crop cultivars'. *Plant Biotechnology Journal* 6: 2–12.

Meyer K (1990). *Pundits, Poets, and Wits.* New York: Oxford University Press p xxxvii.

Moore P (2012). 'Greenpeace's crime against humanity'. http://www.climatedepot.com/a/17217/Former-Greenpeace-cofounder-Dr-Patrick-Moore-rips-Greenpeace-Crimes-Against-Humanity

Mundree SG, Whittaker A, Thomson JA and Farrant JM (2000). 'An aldose reductase homologue from the resurrection plant *Xerophyta viscosa* (Baker)'. *Planta* 211: 693–700.

NIH Guidelines for Recombinant DNA Research.

Nordling L (2010). 'Uganda prepared to plant transgenic bananas'. *Nature* (published online 1 October 2010) www.nature.com/ news/2010/101001/f ull/news.2020.509.html

Olembo N (2008). Intellectual property rights policy. In SD Omamo and K von Grebmer (eds) *Biotechnology, Agriculture, and Food Security in Southern Africa*. Washington, DC: International Food Policy Research Institute p 173–186.

Paarlberg R (2008). *Starved for Science: How Biotechnology is Being Kept Out of Africa*. Cambridge, Massachusetts and London, England: Harvard University Press.

Paarlberg R (2012). (in press).

Paine JA, Shipton CA, Chaggar S, Howells RM, Kennedy MJ, Vernon G, Wright SY and Hinchliffe E et al. (2005). 'Improving the nutritional value of Golden Rice through increased pro-vitamin A content'. *Nature Biotechnology* 23: 482–487.

Phillips McDougall (2011). 'The cost and time involved in the discovery, development and authorisation of a new plant biotechnology derived trait'. A consultancy study for CropLime International.

Potrykus I (2001). Golden rice and beyond. *Plant Physiology* 125: 1157–1161.

Robinson J (2009). *Bluestockings*. London: Penguin Books.

Rome Declaration (1996). United Nations Food and Agriculture Organisation. http://www.fao.org/docrep/003/w3613e/w3613e.htm

Royal Society (2003). 'Where is the evidence that GM Foods are inherently unsafe, asks Royal Society'. Press Release 8 May 2003.

Science in Africa http://www.scienceinafrica.co.za/nerica.htm (accessed 29 October 2011).

Sehnal F and J Drobnik J (2009). Genetically modified crops, EU regulations and research experience from the Czech Republic © Biology Centre of the Academy of Sciences of the Czech Republic, vvi, 2009, Praha. http://www.ask-force.org/web/ Regulation/Sehnal-Drobnik-White-Book-2009.pdf (accessed 13 March 2013).

Shackleton L (1997). *Eve and the tree of knowledge*. Scientech.

Shepherd DN, Mangwende T, Martin DP, Bezuidenhout M, Kloppers FJ, Carolissen CH, Munjane AL, Rybicki EP and Thomson JA (2007). 'Maize streak virus-resistant transgenic maize: a first for Africa'. *Plant Biotechnology Journal* 5: 759–767.

Shepherd DN, Martin DP, Van der Walt E, Dent K, Varsani A and Rybicki EP (2010). 'Maize streak virus: an old and complex 'emerging' pathogen'. *Molecular Plant Pathology* 11: 1–12.

Singh B (2003). 'Improving the production and utilization of cowpea as food and fodder'. *Field Crops Research* 84:150–169.

Sinha G (2007). 'GM technology develops in the developing world'. *Science* 315: 182–183.

Snell C, Bernheim A, Bergé J-B, Kuntz M, Pascal G, Paris A and Ricroch AE (2011). 'Assessment of the health impact of GM plant diets in long-term and multigenerational animal feeding trials: a literature review'. *Food and Chemical Toxicology* 50: 1134–1148.

Southern African Confederation of Agricultural Unions (SACAU) (2011). 'Genetically modified organisms (GMOs) in agricultural development: policy statement'. http://www.sacau.org.

Stock G (2003). *Redesigning Humans: Choosing Our Genes, Changing Our Future*. Harcourt, USA: Houghton Mifflin.

Thomson JA (1977). '*E. coli* phosphofructokinase synthesized in vitro from a ColE1 hybrid plasmid'. *Gene* 1: 347–356.

Thomson JA (2001). 'Horizontal transfer of DNA from GM crops to bacteria and to mammalian cells'. *Journal of Food Science* 66: 188–193.

Thomson JA (2002). *Genes for Africa*. South Africa: UCT Press.

Thomson JA (2006). *GM Crops: The Impact and the Potential*. Australia: CSIRO Publishing.

Thomson JA (2006). *Seeds for the Future*. USA: Cornell University Press.

Thomson JA, Shepherd DN and Mignouna HD (2010). 'Developments in agricultural biotechnology in sub-Saharan Africa'. *AgBioForum* 13: 314–319. http://www.agbioforum.org/

Tokar B (2001). *Redesigning Life? The Worldwide Challenge to Genetic Engineering*. UK: Zed Books Ltd.

US Office of Science and Technology Policy (1986). 'Coordinated framework for regulation of biotechnology'. *Federal Register* 51: 23302–23350.

Van Montagu M (2011). 'It is a long way to GM agriculture'. *Annual Reviews of Plant Biology* 62: 1.1–1.23.

Van Zyl A (2011). 'Innovation in South Africa—the role of the Technological Innovation Agency'. *South African Journal of Science* 107: 1–2.

Varsani A, Shepherd DN, Monjane AL, Owor BE, Erdmann JB, Rybicki EP, Peterschmitt M, Briddon RW, Markham PG, Oluwafemi S, Windram OP, Lefeuvre P, Lett J-M and Martin DP (2008). 'Recombination, decreased host specificity and increased mobility may have driven the emergence of maize streak virus as an agricultural pathogen'. *Journal of General Virolology* 89: 2063–2074.

Visser D and Schoeman AS (2004). 'Flight activity patterns of the potato tuber moth, *Phthorimaea operculella* (Lepidoptera: Gelechiidae)'. *African Entomology* 12(1):133–137.

Vitale JD, Vognan G, Ouattarra M, and Traore O (2011). 'The commercial application of GMO Crops in Africa: Burkina Faso's decade of experience with *Bt* cotton'. *AgBioForum* 13(4): Article 5.

Wambugu F (2001). Modifying Africa: how biotechnology can benefit the poor and the hungry, a case study from Kenya. Nairobi, Kenya.

WHO (1987). *Principles for the Safety Assessment of Food Additives and Contaminants in Food, Environmental Health Criteria 70.* Geneva: World Health Organization.

WHO (1991). *Strategies for Assessing the Safety of Foods Produced by Biotechnology.* Report of a Joint FAO/WHO Consultation. Geneva: World Health Organization.

www.absfafrica.org (accessed 10 October 2011).

Index